The Crumbs of Creation

Trace elements in history, medicine, industry, crime and folklore

The Crumbs of Creation

Trace elements in history, medicine, industry, crime and folklore

John Lenihan

Department of Nursing Studies,
University of Glasgow

Adam Hilger, Bristol and Philadelphia

British Library Cataloguing in Publication Data
Lenihan, John, *1918–*
    The crumbs of creation.
    1. Man. Trace elements
    I. Title
    612′.3924

    ISBN 0-85274-390-4

US Library of Congress Cataloging-in-Publication Data
Lenihan, J. M. A.
    The crumbs of creation.

    Bibliography; p.157
    Includes index.
    1. Trace elements in the body. 2. Trace elements—Toxicology.
3. Trace elements—Health effects. I. Title.
QP534.L46   1988   615.9   88-6662

    ISBN 0-85274-390-4

Published under the Adam Hilger imprint by IOP Publishing Ltd
Techno House, Redcliffe Way, Bristol BS1 6NX, England
242 Cherry Street, Philadelphia, PA 19106, USA

Printed in Great Britain at The Bath Press, Avon

Contents

Foreword

This book had its origin in a course of lectures which I gave while serving as Regents' Professor in the Chemistry Department of the University of California, Irvine—a stimulating environment which was made all the more agreeable by the kindness and courtesy accorded to me by both staff and students. Mr Michael Purcell suggested that the lectures should be expanded for publication and Miss Nancy Truglio allowed me to make use of the excellent notes which she wrote during the course. I owe a particular debt of gratitude to Professor Vincent Guinn, whose friendship and collaboration, extending over nearly three decades, have much enhanced my knowledge and understanding of trace element chemistry in its widest context.

I have, for an even longer spell of years, enjoyed involvement—more often as a quartermaster than an active participant—in the imaginative and fruitful programme directed by Professor Hamilton Smith of the Department of Forensic Medicine and Science of the University of Glasgow. In the pages which follow I have drawn freely on the published and unpublished research of Professor Smith and of others who have worked with us—notably Dr Ian Dale, Dr David Lyon and Dr Janet Warren.

Chapter 5 contains material which originally appeared in *The Practitioner*. I am grateful to the editor and proprietors of that eminent journal for permission to reproduce it here.

The tasks of gathering and classifying information from a multitude of sources and the drafting of text have made heavy demands on the patience and devotion of many colleagues—especially Mrs May Buchanan, Mr Douglas Craig, Mrs Margaret Tait, Miss Anne Wotherspoon and Mrs Ellen Woodards. I have also enjoyed the privilege of access to the incomparable riches of the Glasgow University Library and to the ever helpful expertise of its staff.

At the University of California, Santa Barbara, my search for information on Drake's Plate (Chapter 7) succeeded through the help of Dr Joseph Boise, University Librarian, and Mr Christian Brun, Head of the Department of Special Collections.

Dr Alan Richardson and Mrs Jean Robertson have read the whole of the text; Professor Guinn and Mr Herbert Soulsby read some of the chapters. I am grateful to these friends for their thoughtful comments.

During the years when this book was in preparation I have been supported by the encouragement and practical help of Mrs Lena Fleming, Professor Agnes Jarvis, Miss Deborah Rogers, Mrs Patricia White and—most of all—my wife.

Several of my earlier writings were illustrated by Mr Jack Fleming, whose witty and elegant drawings added a new dimension to the scientific content. Our collaboration was, to my profound sorrow, ended by his death in 1986. Drawings which he had completed enrich the text of this book—and remind us of what might have been.

I appreciate the care and skill with which the Adam Hilger staff have handled the preparation of this book; in particular, Neville Hankins (commissioning editor) and Neil Robertson (desk editor).

John Lenihan
Glasgow
January 1988

Introduction

God, according to Voltaire, is on the side of the big battalions. The Universe certainly seems to demonstrate this strategy. Almost everywhere the environment is dominated by just a few chemical elements—hydrogen and helium in outer space, oxygen and silicon in the Earth's crust, oxygen and hydrogen in living matter. At the other end of the scale, many elements are found in extremely small concentrations, measured in parts per million (PPM), parts per billion (PPB) or even less. These are the trace elements.

The concept originated during the 19th century. Analytical chemists, when reporting on a sample, would state the relative proportions of its constituents. Any which were present at levels sufficient to be identified but not measured were reported as 'traces'. The first recorded example of this usage is attributed to Faraday. He wrote in 1827 of a process occurring in a blast furnace 'leaving scarcely a trace of slag'. He may have been thinking of a trace as a streak, which is how a small amount of slag would appear on molten metal, but the word was useful in the rapidly-developing science of analytical chemistry—and in other fields also.

In 1876 Mr W E Gladstone, writing (in the *Contemporary Review*) on varieties of religious belief, offered a new simile: 'Like a chemist who, in a testing analysis . . . if he finds something behind so minute as to refuse any quantitative estimate, calls it by the name of "trace".'

As the result of advances in analytical chemistry, particularly during the past 40 years, virtually all of the elements formerly designated as traces can now be accurately measured. Some appear to be of little significance in living systems but others, with which this book is concerned, demonstrate importance out of proportion to their meagre abundance.

We define a trace element as one which is present to a very small extent in a particular environment and which either

(a) by its presence gives significant information,
or
(b) by the expression of its chemical and physical properties exerts a distinctive influence (adverse or advantageous) on its environment.

In the pages which follow, the role of trace elements will be reviewed in relation to several environments.

(i) The geological environment is, for our purpose, the crust, oceans and atmosphere of the Earth. We shall discuss the contribution of trace element analysis to understanding of the end of the dinosaurs, to exploration for minerals, to environmental pollution and to the role of water in human health.

(ii) The biological environment includes all living matter. We shall examine the role of trace elements in plant, animal and human nutrition and in various occupational and environmental hazards to human health.

(iii) In the context of the social environment, we shall review the contribution of trace element studies to the detection of crime—contemporary and historical—and to scholarly activity in art and archaeology.

In trying to identify, among the 90 naturally-occurring elements listed in appendix 1, those which we shall study in this book as trace elements, we can start by excluding 11 which are the most abundant in the tissues of man and the higher animals and plants and may therefore be regarded as bulk elements. They are: hydrogen, carbon, nitrogen, oxygen, sodium, magnesium, sulphur, chlorine, phosphorus, potassium and calcium. Of the remaining 79 elements, we can exclude the six noble gases. These (helium, neon, argon, krypton, xenon and radon) were for many years regarded as completely inert; it is now known that some of them do form compounds—but none of biological relevance. We can also exclude the seven heavy elements (polonium, francium, radium, actinium, thorium, protactinium and uranium) which are invariably radioactive, as they have no stable isotopes.

We are left with 66 elements, all of which are found in the earth's crust and are therefore present in animal and plant tissues—though not always at levels within the reach of existing analytical techniques. These might all be regarded as trace elements. In this book we shall be concerned only with those which are essential for the life of animals (including man) or plants, and a few more which are interesting because of their toxic effects. These restrictions reduce the list to 19: boron, fluorine, silicon, vanadium, chromium, manganese, iron, cobalt, nickel, copper, zinc, arsenic, selenium, molybdenum, cadmium, tin, iodine, mercury and lead.

The study of trace elements has been transformed by great advances in the techniques of analytical chemistry during the past 40 years. The first breakthrough was achieved by the development of neutron activation analysis during and after the Second World War. Many elements become radioactive when exposed to neutron bombardment, usually inside a nuclear reactor. Assay of the radioactivity induced in a sample, which is highly characteristic of the elements contributing to it, forms the basis of an extraordinarily sensitive analytical procedure, applicable to most of the trace elements mentioned in this book. Other techniques, offering comparable or greater sensitivity for many trace elements, have become available more recently and have been used in some of the work described here.

CHAPTER 1

Hair and History

A KING KILLED BY CHEMISTRY?

The King was not happy. As 1685 began, his years of exile and adversity were long gone, he was politically secure, his kingdom was peaceful and prosperous, he had both the inclination and the means to live in extravagant style—but his health was declining. Prevented by an attack of gout from taking his daily walk in St James's Park, he spent more time than usual in his laboratory, distilling mercury (a process favoured by the alchemists in their efforts to make gold) and otherwise dabbling in the sciences of which he was such a notable patron. Charles II is revered today as the founder of the Royal Society—the world's most illustrious learned society—but in his own age he was criticised for spending too much time in his laboratory and not enough in the councils of state.

On Sunday 1 February 1685 his old exuberance returned. He passed the day in his palace at Whitehall in the company of revellers and gamblers, carousing with three of his mistresses and generally enjoying the robust pleasures for which his court was notorious. On the following morning the courtiers noticed that his speech was indistinct and his mind confused. Suddenly he collapsed. His physicians—fourteen of them—argued among themselves while administering unpleasant and useless remedies.

On 6 February the King died. Apart from the inevitable rumours of poison, the cause of his last illness remained obscure for nearly three centuries. In 1961, two American physicians reviewed the evidence and produced a diagnosis of kidney failure, attributable to mercury poisoning; this was quite consistent with the King's well-authenticated history of exposure to mercury vapour in his laboratory.

His physicians argued among themselves.

Seeking further support for their diagnosis, they suggested that analysis of the King's hair might show whether he had absorbed a significant amount of mercury. In 1966, after a broadcast appeal, a viewer in Wales donated a small sample for examination by Dr (now Professor) Hamilton Smith in the Department of Forensic Medicine at Glasgow University.

This relic came from a lock of hair, mounted on a card bearing the words:

'This lock of hair was taken from the head of King Charles the 2nd, by the mother of Sir John Jennings Kt, and given to Miss Steele of Bromley by Phillip Jennings Esq. Nephew to the Admiral Sir John Jennings above said 1705'.

The King's hair was found, by neutron activation analysis, to contain 54 PPM of mercury. This is a very high value (about 20 times greater than today's normal level) and gives some support to the long-range diagnosis of the American physicians—although, since it is not known when the lock was cut from the King's head, it is not possible to be confident that he died of mercury poisoning. The investigation does, however, illustrate the use of hair analysis in the study of historical problems. To appreciate the value of this technique we need to know more about the physical and chemical properties of hair.

STRUCTURE AND PROPERTIES OF HAIR

Hair is a fascinating though undervalued material, offering much of interest to workers in medicine, forensic science, environmental hazard control and even history. In these disciplines its usefulness, which cannot always be exploited without difficulty or controversy, depends on several factors:

(1) it records contamination from inside the body (since it is a pathway of excretion for many metals) and from outside, since it traps metallic vapours and dust.
(2) Growing at a fairly uniform rate of about a centimetre a month, it acts as an integrating dosemeter, with a built-in time scale. Blood and urine, though widely used as indicators of exposure to toxic substances, are of limited value for this purpose since they record only recent exposure.
(3) It can be collected quickly and painlessly without specialised skill or equipment and (in contrast to most other tissues) can be stored and transported cheaply without deterioration.
(4) Many trace elements are found at higher concentrations in hair than in other tissues (table 1.1).
(5) It can be analysed with great sensitivity by techniques which are widely available.
(6) It is very durable. Samples from as far back as the 4th century show little change in physical characteristics.

To understand how these properties come about, we need to know something of the physical and chemical structure of the material. Hair, like skin and nail, is composed largely of keratin (a tough protein) and may be regarded as a solid secretion of the skin. Its formation from materials in soft tissue, blood and other body fluids begins in the follicles which cover most of the body. In man there are about 100 000 on the scalp, 10 000 to 20 000 on the face and up to a million on the rest of the body.

A hair consists of four parts, not of equal value or interest to the analyst. The root is embedded in the follicle. The shaft, which emerges through an opening in the skin, has three parts. The cuticle is a structure of overlapping scales, giving protection from mechanical and chemical insults. The cortex, a hollow cylinder which makes up most of the hair, is composed of keratin fibres. The medulla is a central cavity, usually absent in fine hair and not always continuous in coarse hair, as on the scalp. When present, it is filled with fibre-walled cells (which contribute stiffness) separated by air spaces.

Table 1.1 Typical trace element concentration, PPM.

Element	Whole body	Blood	Hair
aluminium	1	0.4	5
arsenic	0.2	0.005	0.5
gold	0.01	0.0001	0.1
bromine	4	5	30
cadmium	0.7	0.005	1
chromium	0.1	0.1	1
copper	1	1	15
mercury	0.2	0.01	2
magnesium	0.3	0.3	40
manganese	0.2	0.05	0.3
nickel	0.01	0.03	3
lead	2	0.2	20
selenium	0.2	0.2	20
titanium	0.2	0.05	4
zinc	330	7	150

The colour of the hair is influenced by the chemical composition of the melanin, a pigment which occurs in the medulla and, to a lesser extent, in the cortex. Hair becomes whiter in old age, partly through loss of pigment and also because the cells in the medulla shrink; the intervening air bubbles then become more numerous and the hair reflects more light.

In man (as in the cat and the guinea pig, though not in most other animals) each hair grows independently of its neighbours; that is why we don't moult. At any time, about 90 percent of the hairs on a human head are growing. The remainder, except for a few at an intermediate stage, are resting. The growth phase lasts for about 900 days and is followed by the resting phase which continues for about 100 days. After that, one of two processes occurs; the root may shrink, allowing the hair to fall out, or a fresh growth cycle may begin.

The major chemical constituents of hair, as of every other tissue, are hydrogen, carbon, nitrogen and oxygen. The sulphur content is unusually high—about four percent. The five elements mentioned make up well over 99 percent of the mass. About 60 other elements have been found in hair; no doubt diligent search using sufficiently sensitive techniques would reveal more.

The presence of some elements, such as copper, zinc and phosphorus, which are known to be essential elsewhere in the body, is not surprising—but the value of hair as an analytical sample is usually related to the study of the less common elements.

The distinctive value of hair analysis depends on the fact that its chemical constituents, unlike those of most other tissues, are not in a state of flux, characterised by frenzied breakdown and rebuilding of molecules. When amino acids in the follicle are changed into keratin, the hair which emerges is no longer a living tissue. The appearance of growth is given by the extrusion of fresh material from the follicle, rather like toothpaste from a tube.

During the keratinisation process materials from the blood or other body fluids may be incorporated into the hair protein. The attachment of metals to protein molecules is a familiar process, sometimes beneficial (as when metals are associated with enzymes as structural elements or catalysts) and sometimes harmful, as when an essential metal is displaced from a protein molecule by a toxic interloper. The incorporation of elements such as mercury and arsenic into keratin does no harm (since the hair does not engage in metabolic activity) but is useful to the extent that subsequent chemical analysis gives information about materials present in the immediate environment at the time of keratinisation. Metals may also be deposited on (or incorporated in) hair by external contamination; there are many possible sources, including dust, vapours, sweat and cosmetics.

The ability of hair to concentrate many metals (table 1.1) is attributable to its high protein content—and in particular to the abundance of cystine, an amino acid containing in each molecule a disulphide (S-S) group to which metals readily become attached. Other amino acids in hair protein contain sulphydryl (SH) groups, which also have an affinity for metals.

Why do we need to analyse hair? The basic reason is that information the amount of a particular element present in the body, or in one part of the body, is sometimes useful in the study and management of illness or of environmental hazard. For major elements with well understood metabolic roles, isotope dilution tests using radioactive tracers are often serviceable. But when an appreciable amount of an element is stored at a site which is not amenable to exchange with an added radioactive tracer (for example, lead in bone or cadmium in kidney) the body burden has to be inferred from measurements on samples which are readily available. Blood and urine are traditionally used in this way, but hair often gives more useful information.

For potentially toxic elements the cumulative intake over a period of time is significant. Hair analysis helps in two ways. First, the concentration of an element in a hair is related to the average uptake into the blood (which is usually proportional to the intake into the body) over a period of time which depends on the length of the hair examined; as already mentioned, human head hair grows at about a centimetre a month. Secondly, the time variation of uptake can be inferred by analysing small sections of a single hair; a one millimetre length corresponds to about three days' growth. In male subjects the analysis of daily samples of beard hair (taken with an electric razor) gives information on the rate of uptake or, after exposure has ceased, on the rate at which the body burden is decreasing.

INTERPRETING THE RESULTS

In interpreting the findings from hair analysis two questions need to be kept in mind: (a) is the hair sample representative? and (b) has the element in question reached the hair by the internal route (through the blood) or by an external route involving environmental pollution or cosmetic preparations.

Since roughly 90 percent of the hairs on a human head are, at any time, in the growth phase and 10 percent in the resting phase, a representative sample should contain the two types in the same proportion. It is therefore advisable to take a bundle of at least 20 hairs. This precaution is necessary also because, for many trace elements, the concentration varies quite widely among individual hairs. When it is not possible to obtain a sample of the desirable size (for example in historical studies or in the examination of stray hairs from the scene of a crime) hair in the growing phase should be sought; the two phases are easily identified by the differing shapes of the hair root.

Studies involving groups of people, for example in the investigation of environmental hazards, are subject to further uncertainty because the concentration in hair of many trace elements varies quite widely throughout the population—even among people exposed to closely similar diets and environments. A study made in Holland in 1965 involved the analysis of hair samples from 16 Trappist monks, whose diet and environment were identical. Concentrations of zinc, bromine and gold showed little variation from one individual to another, but considerable differences were found for arsenic, chromium and mercury. Other work has shown that, in most people, levels in hair of copper and zinc do not change much over long periods—but the concentrations of other trace elements can fluctuate markedly over periods of a few weeks or less.

External and internal contamination of hair may occur at measurable levels in many industrial environments as well as from cosmetic preparations. Selenium occurs in anti-dandruff lotions, thallium in hair removers, zirconium in deodorants and magnesium in talcum powders.

It is sometimes suggested that contamination by sweat or sebum (the greasy substance which forms a thin protective layer over the skin, particularly on the scalp and face) may invalidate conclusions drawn from hair analysis. This objection may be significant in regard to sectional analysis of hair (to establish the time course of exposure) but is less important in relation to the analysis of whole hairs, since sweat and sebum are derived ultimately from the same sources as hair.

LOST—AND GAINED—IN THE WASH

It might be thought that external contamination could be removed by washing the hair. This is a contentious issue. Some washing procedures, involving detergents or organic solvents, remove trace elements from the interior of the hair; others add them. In these respects the same washing procedures sometimes give different results when used by different investigators. Agreement of a standard method of washing would be useful but has not yet been achieved. It is of course sometimes advisable to analyse hair without any preliminary treatment—for example when studying external contaminants which may also be inhaled or swallowed. Analysis of unwashed samples (taken at the same time) of head hair and body hair allows some discrimination between external and internal contamination.

Washing of hair in the hope of removing external contamination can produce strange results. During a large-scale survey of hair arsenic levels in normal subjects, conducted in Glasgow in 1956, it was found that a number of female laboratory workers showed very high values, up to 42 PPM—nearly 100 times the normal level in female hair. It was thought that external contamination might have come from laboratory chemicals or from materials used in the adjoining animal house. Further samples obtained from the workers whose hair showed abnormally high arsenic levels were therefore cleaned by soaking overnight in detergent solution. The arsenic levels were little changed by this treatment. As a control experiment, samples of hair with normal arsenic content were cleaned in the same way. Analysis of these samples showed arsenic levels up to 23 PPM.

It was then thought, as no other explanation seemed possible, that the detergent must have contained arsenic. Analysis of a sample taken from the laboratory supply showed an arsenic content of 74 PPM. A detergent sold by the same manufacturer under a different brand name contained 59 PPM. Other detergents, some liquid and some powdered, were found to contain arsenic at concentrations up to 8 PPM.

The origin of the arsenic was found without much difficulty. Many detergents, both solid and liquid, contain components made by the action of sulphuric acid on hydrocarbons, alcohols or other organic materials. The arsenic poisoning outbreak of 1900 (p 111) was attributed to contamination of sulphuric acid made by the lead chamber process. In this operation sulphur is obtained by burning pyrites (iron sulphide) in which arsenic is invariably present as an impurity.

When the detergent manufacturers confirmed that they used sulphuric acid made in the lead chamber process the mystery was solved—or almost so. It remained to explain the curious findings which had initiated the enquiry. The hair samples washed in the laboratory had absorbed arsenic from the detergent solution applied in the hope of cleaning them. The high levels of arsenic found in the untreated samples from female laboratory workers were attributed to the practice of washing the hair with undiluted detergents borrowed from the laboratory and failing to rinse adequately.

Appropriate advice was given but disciplinary action was not thought to be necessary. The manufacturers took prompt and effective action to purify their products by finding a source of relatively uncontaminated sulphuric acid. The arsenic content of the offending brands of detergent fell rapidly; samples bought a year later showed arsenic levels under 2 PPM.

WAS NEWTON MAD?

Laboratory workers of earlier times were often vulnerable to internal and external contamination—none more so than the alchemists. It is fashionable to dismiss as folly, motivated by greed, the quest for the Philosopher's Stone, which would cure all the ills of man and convert base metals into gold. Alchemy was certainly pursued by many charlatans, but fascinated many eminent enthusiasts too. Charles II was a dilettante, but the craft attracted serious scholars, including John Locke the philosopher, Robert Boyle and Isaac Newton, England's greatest scientist.

For 30 years or more, beginning in 1669, Newton experimented with mercury, arsenic, lead and antimony in a variety of procedures from the alchemists' repertoire. Sometimes he tasted the products of his experiments. He made an amalgam by rubbing gold dust into mercury, in the palm of his hand, and on many occasions inhaled mercury vapour. In 1692 and 1693 he suffered a serious illness, characterised by insomnia, loss of appetite, irritability towards his friends and, for two short periods, delusions of persecution. These symptoms are consistent with a diagnosis of depressive illness—but have been attributed by some biographers to temporary insanity. Two reports published in 1979 suggested mercury poisoning as the cause.

In one of these studies, four samples of Newton's hair were analysed for several trace elements. One sample was of doubtful authenticity; two others were reported as showing abnormally high concentrations of mercury—54 and 197 PPM. The significance of these findings is uncertain, for two reasons:

(1) The analysis of the two samples indicated sodium content of twice the normal level and chlorine at five times the normal level, suggesting that the hair had been grossly contaminated.

(2) The date when the samples were taken is not known. The authors of the report surmised that they had been obtained immediately after Newton's death in 1727. It would, of course, be impossible to infer anything at all about an illness which occurred in 1692/3 from the analysis of hair taken 34 years later.

It is very probable that Newton absorbed an undesirable quantity of mercury during his alchemical experiments, but there is not enough evidence to link this exposure with his illness of 1692/3.

ANCIENT HAIR

The durability of hair was well illustrated during an investigation conducted by archaeologists and chemists in 1971.

A young man with waist-length hair tied in a pigtail would probably not be considered eccentric today; nor would he have been in Durnovaria (now Dorchester) an elegant and prosperous outpost of the Roman Empire. Plaited hair might have been considered effeminate in Rome—but in 4th century Britain the Celtic hairstyles of pre-Roman times still survived.

The lead coffin which was uncovered (and damaged) in 1971, by a mechanical excavator preparing foundations for a new Government building in Dorchester, provided archaeologists with a great amount of information. Some of the bones and other contents were destroyed, but enough remained to establish that the corpse had been that of a man in his twenties. He had at some time suffered a broken collarbone and several broken ribs. His body had been wrapped in a fine linen shroud, an indication of affluence which was supported by the excellent workmanship of the coffin.

The most remarkable find was the young man's hair, which was virtually complete and little damaged after sixteen centuries underground. It was loosely plaited; a three-ply pigtail, 28 cm long, had been cut off after death. After ultrasonic cleaning in a dilute solution of detergent the hair was very similar, in feel and in appearance, to modern hair. These impression were confirmed by a battery of tests on its physical and chemical composition.

A small sample was analysed by Dr Ian Dale in Glasgow. The concentrations of arsenic (0.22 PPM), cadmium (0.45), bromine (34.7) and copper (42.0) were all within the range shown by modern hair. The presence of lead (760 PPM) and antimony (12.6) at about 20 times the levels common today was attributed to contamination from the material of the coffin. Mercury, found at 6.1 PPM (about twice the level common today) may have come from cosmetics or medicines.

Mercury was certainly known to the physicians of Roman times (p 88). The sulphide was used as a medicine and as a colouring agent in cosmetics. The ointment made by rubbing mercury and grease together survived in the pharmacists' prescription books well into the present century.

A GRUESOME CUSTOM

Evidence of the liberal use of mercury—possibly in ointment—was found in hair from a 13th century burial site in Dumfries-shire. A small lead casket uncovered during excavations at Sweetheart Abbey contained a severed head, giving evidence of a rather gruesome custom. When the demand for burial space in consecrated ground exceeded the supply, the head was sometimes buried in the church and the body elsewhere. The casket found at Sweetheart Abbey contained a full head of hair, little changed by the passage of time. The mercury content was 42 PPM.

THE LITTLE PRINCESS

Trace element analysis of hair, conducted by Professor Hamilton Smith and his colleagues in Glasgow, has produced biographical information about some historical personages.

The medical history of an early member of the British royal family was reopened after the discovery, on a London building site in 1964, of a small lead coffin containing the remains of Ann Mowbray. She was the child wife of Richard, Duke of York, the younger of the Princes in the Tower. The circumstances of her death in 1481, at the age of eight, were uncertain. Her hair was well preserved and a sample was sent to Glasgow for analysis by Professor Smith.

It contained arsenic at 3.3 PMM and mercury at 9.1 PPM. These values are higher than would be expected today, but not in themselves high enough to support any suggestion of poisoning. Since analysis of whole hairs gives information only about total uptake over several months, a further test for arsenic was made by cutting one hair into 15 pieces, each one centimetre long. The arsenic levels in successive pieces, starting from the head end, were (in PPM):

11.0, 4.1, 3.7, 12.6, 3.4, 7.2, 0.9, 1.8, 2.6, 0.9, 2.6, <0.1, <0.1, 3.5, 11.3.

Here again the levels are not high enough to suggest that toxic amounts had been swallowed. In the 15th century—and for long afterwards—arsenic was commonly used in medicine and in the adulteration of food. The distribution of arsenic in the hair suggests intermittent administration, almost certainly from such sources.

When the hair analysis and other tests were completed the remains of little Ann were affectionately re-interred in Westminster Abbey, where she had been married with great ceremony nearly five centuries earlier.

THE CASE OF THE POISONED EMPEROR

'I die before my time, murdered by the English oligarchy and their hired assassin'.

Napoleon's will, written in his last days, at the end of his five years of exile on St Helena, is a long and spirited document. The Emperor's health had deteriorated since he arrived on the island in 1816. He was convinced that the Governor, Sir Hudson Lowe, had been instructed to kill him, either by poison or by the sword.

There is no doubt that Napoleon's health was impaired during his imprisonment. The island was, like many tropical territories at that time, an unhealthy place, but Hudson Lowe had little concern for his prisoner's health. Dr Barry O'Meara, a naval surgeon who reported that Napoleon was suffering from tropical hepatitis (an unpleasant disease now known as amoebiasis) and that his health would be further endangered if he remained on St Helena, was sent back to England in disgrace and dismissed from the Navy. His successor, John Stokoe, who agreed with O'Meara's diagnosis, was tried by court martial and also dismissed from the service. He was followed by Francesco Antommarchi, a Corsican, who was an anatomist rather than a physician. Napoleon appraised his clinical competence in forthright terms:

'I would give him my horse to dissect, but would not trust him with the cure of my own foot'.

The *post-mortem* examination conducted by Antommarchi on the day after Napoleon's death in 1821 was observed by no fewer than 17 spectators—doctors, Army officers and servants. A variety of opinions emerged. Antommarchi reported a cancerous ulcer of the stomach, as well as a diseased liver. The British doctors diagnosed cancer of the stomach; their opinion was acceptable to Hudson Lowe, who resented any suggestion that Napoleon's life might have been shortened by neglect of the advice given by O'Meara and Stokoe. Antommarchi may have been under some pressure when writing his report; after he returned to Europe, he asserted that the diagnosis of cancer had been made only by the British doctors, and that in his opinion Napoleon had died of hepatitis, acquired on the island.

Napoleon's last illness has been a subject of speculation and controversy ever since. Fresh interest was aroused by the appearance in 1962 of Dr Sten Forshufvud's book, provocatively titled: *Who killed Napoleon?* Dr Forshufvud, a Swedish dentist and an earnest student of Napoleonic literature, reviewed the memoirs of the Emperor's medical attendants and of many others who were present on St Helena between 1816 and 1821. He concluded that Napoleon had been poisoned, in a rather complicated way involving three procedures:

(1) administration of arsenic in toxic (but not immediately lethal) doses over a period of years;
(2) administration of a large dose of tartar emetic (potassium antimonyl tartrate) to induce vomiting in order to remove any traces of arsenic from the stomach, where it might be detected after death;
(3) administration of a large dose of calomel (mercurous chloride), along with a substance which would convert it, in the stomach, to the more poisonous corrosive sublimate (mercuric chloride) which would quickly kill the victim.

Knowing of Professor Hamilton Smith's reputation as the foremost authority on the estimation of arsenic in hair, Dr Forshufvud sent him two small samples of Napoleon's hair, telling him only that they had come from a subject believed to have been exposed to arsenic.

Professor Smith, who was well accustomed to studying such material, made the necessary tests and reported arsenic levels of 10.38 PPM in one sample and 10.53 PPM in the other. Dr Forshufvud then disclosed the origin of the samples. The first had been among the belongings of Louis Marchand, Napoleon's chief valet on St Helena. It was in a packet inscribed in Marchand's writing: 'Cheveux de L'Empereur'. The packet bore no date, but Dr Forshufvud believed that the hair had probably been cut from Napoleon's head shortly after his death. The second sample had belonged to Jean Baptiste Isabey, Court Painter to Napoleon and had been given to him in 1806. Both samples had latterly been in the custody of Commandant Henri Lachouque, a distinguished Napoleonic scholar.

Professor Smith's findings, prefaced by a summary of Dr Forshufvud's historical research, were published in *Nature* (a British scientific weekly) in October, 1961 and were widely reported in the media. Three weeks later, Mr Clifford Frey, a Swiss textile manufacturer, came to Glasgow with a family heirloom in the form of a bundle of Napoleon's hairs. This precious relic, in a small envelope inscribed: 'Cheveux de L'immortel Empereur Napoléon', was enclosed in a larger envelope containing a letter written in 1838 by Abram Noverraz (one of Napoleon's valets) to a friend in whose family it remained until acquired by Mr Frey's father.

The letter explained that the hair had been taken from Napoleon's head on 6 May, 1821—the day after his death. Three of these hairs were analysed and found to contain 18.0, 10.2 and 12.3 PPM of arsenic. These values are, of course, well above the levels that would be expected in unexposed persons today—but a more valid assessment of their significance can be made by reference to normal arsenic levels in human hair in Napoleon's time. Some information is available on this point. Hair samples from persons who died between 1790 and 1849 have been analysed in Glasgow; they show an average (geometric mean) level of 3.8 PPM.

During the first few months of his exile, Napoleon was accommodated in the house of Mr William Balcombe, Superintendent of the East India Company (owners of St Helena before the British government took it over). Mr Balcombe's great-granddaughter (Dame Mabel Balcombe–Brookes) supplied some hairs from a locket given by Napoleon to Mr Balcombe's daughter Betsy in 1818. Two of these hairs were analysed and found to contain 21.2 and 7.5 PPM of arsenic.

Admiral Sir Pulteney Malcolm, commander of the naval force guarding St Helena, paid his respects to the Emperor before returning to England in 1817. He received a lock of hair, from which a sample was donated by one of his descendants. The arsenic content was 3.2 PPM.

Dame Mabel Balcombe–Brookes donated another sample, consisting of four short hairs (2-4 cm in length) and a few odd scraps, from a lock given on 14 January 1816 to Commander John Theed, a British naval officer whose ship (HMS Leveret) delivered mail to St Helena. The arsenic content of the four hairs which made up most of the sample was 39.1 PPM. Since much of Napoleon's hair was—on the evidence of contemporary drawings and paintings—longer than 4 cm, there is no reason why the sample under consideration should have been cut close to the scalp. Its arsenic content may therefore reflect exposure incurred several months earlier than January 1816—that is, at a time when Napoleon was still at liberty.

The analysis of the Emperor's hair suggests that the highest intake of arsenic occurred before he reached St Helena. This finding does not support a diagnosis of deliberate poisoning. It is more probable that

Napoleon was given arsenical medicines—which were widely popular in those days—on the advice of his doctors, both before and after his capture in 1815.

DID ROBERT BURNS DIE OF DRINK—OR MERCURY?

'Fame, prompted by priests, yes, countenanced by friends, has promulgated an untruth that Burns died, prematurely died, dissipations's martyr. From personal correct knowledge, I proclaim that Robert Burns died the doctor's martyr'.

The medical history of Robert Burns has been studied with enthusiasm by generations of clinicians, patriotic Scots and armchair pathologists. Though the consensus of expert opinion rests on heart disease of rheumatic origin as the cause of the poet's death, his early biographers, led by James Currie, ascribed his downfall to excessive indulgence in alcohol—a judgment (and a way of life) which still has many supporters in his native country.

John Thomson, whose opinion is quoted above, was a teacher in Dumfries during the 1790s and was among the poet's friends. Later he qualified in medicine in Edinburgh and practised there. His defence of Burn's reputation was contained in a book published in 1844, under the title: *Education: Man's Salvation from Crime, Disease and Starvation; with Appendix vindicating Robert Burns*. The appendix continued:

'The truth stands thus—The physician of Robert Burns believed that his liver was diseased, and placed him under a course of mercury. In those days a mercurial course was indeed a dreadful alternative. I know well that his mercurial course was extremely severe . . . Among the last words I ever heard him speak were, "Well, the doctor has made a finish of it now".'

The physician referred to was Dr William Maxwell. Described by Sir James Crichton-Browne as 'that very haphazard practitioner', he does not appear to have left any record of his patient's symptoms or treatment. It was, however, possible to study the problem a little further when, in 1971, a sample of Burns's hair (donated as a result of a broadcast appeal) was examined in Glasgow by neutron activation analysis. Professor Hamilton Smith and Dr Alistair Leslie reported a mercury level of just over 8 PPM. This is a moderately high value; a typical level today would be 2-3 PPM, attributable partly to agricultural chemicals and other products not known in the 18th century. Mercury was, however, widely used in medicine. It seems that Burns must have taken mercury in some form—no doubt on the prescription of his physician. But a hair level of 8 PPM would not be regarded with concern today; many dentists carry greater amounts without obvious consequences—though some have come to grief (p 81). The date of the sample of Burns's hair is not known; a sample taken near the end of his life might enable Dr Thomson's assertion to be assessed with more confidence.

CLEANER THAN WE WERE?

The study of trace element levels in the hair can provide interesting anecdotes about famous people—and occasional footnotes to the history of their times. Work of this kind also has a more serious purpose, for it can give us uniquely valuable information about the chemical environment of earlier centuries.

In bygone days, the ritual of mourning often included the distribution, to relatives and friends, of small locks of a deceased person's hair. These mementoes, usually documented with the age and date of death, were usually carefully preserved for sentimental reasons. A collection of material of this kind was assembled in Glasgow during the early 1970s, through the help of the Royal College of General Practitioners, the Scottish Record Office and other custodians. Analysis for trace elements gave the results shown in table 1.2

Table 1.2 Trace element levels in hair (geometric mean, PPM).

Date of sample	Arsenic		Cadmium		Copper		Lead		Mercury		Zinc	
	n	PPM	n	PPM	n	PPM	n	PPM	n	PPM	n	PPM
1799–1849	13	3.8	12	0.14	13	16	12	40	13	3.6	13	148
1850–1899	13	3.7	10	0.1	13	18	10	60	13	6.1	13	167
1900–1949	17	0.8	17	0.2	17	25	17	30	17	1.3	17	209

(n = number of samples)

These findings should be interpreted with caution. The older samples came from many parts of Britain, but the modern samples were taken from people in the Glasgow area and may not be representative of a wider environment. It is, however, reasonable to conclude that internal pollution by mercury has not increased during the past two centuries. There has, of course, been a great increase in the use of mercurial compounds in agriculture and other industries, leading to some environmental disasters (p 84), but pollution control has become more effective—and the medical uses of mercury have almost disappeared. Changes in medical practice, stricter supervision of the purity of food, and legislation restricting atmospheric pollution have been responsible for the unmistakable fall in arsenic levels in hair.

Many contemporary problems in environmental health are amenable to study by hair analysis for trace elements.

Table 1.3 Lead in children's hair.

	Water tank and pipes	Number of subjects	Lead in hair PPM, geometric mean
East Kilbride	copper	81	10.2
Glasgow	copper	104	28.9
Glasgow	lead	35	57.7

Susan, a sixth-year pupil in a secondary school in East Kilbride, a new town near Glasgow, was looking for a project of social relevance and chemical interest when she called at an environmental science laboratory which had an active programme of trace element research. Soon she was enrolled as an unpaid part-time assistant in a project examining lead concentrations in children's hair. Earlier work on blood lead levels, in a different laboratory, had looked also at the mental development of children in relation to lead levels in drinking water. The aim of the Glasgow project was to see whether hair analysis could provide an index of internal pollution from drinking water and external pollution from sources in the urban environment. Hair samples were obtained from three groups of children living (a) in East Kilbride, where there is no lead plumbing, (b) in modern Glasgow housing with copper water pipes and tanks and (c) in older Glasgow housing with lead tanks and pipes. Susan organised the collection of samples from her home town and shared in the analysis. The results (table 1.3) gave a striking demonstration of the value of trace element analysis in the study of environmental pollution. Susan is probably the only person to have been a contributor to the eminent medical journal *The Lancet* while still a schoolgirl.

HAIR AND HEALTH

The study of differences in trace element concentrations in hair between normal subjects and patients suffering from particular diseases has been a popular topic of research for many years. This work is done in the hope of improving diagnosis, developing screening tests or illuminating the mechanism of disease. Anomalous concentrations of trace elements have been reported in many diseases, but no widely-accepted diagnostic tests have emerged.

Hair analysis has, in recent years, found a place in Alternative Medicine—both in diagnosis and in treatment. A sample of the patient's hair is sent for analysis, usually to a commercial laboratory. Concentrations of a number of trace elements are measured and interpreted in a report indicating deficiencies or excesses. It is usually recommended by the analytical laboratory that deficiencies should be corrected by oral supplements and excesses by dietary correction or by administration of appropriate chemicals.

These exercises in hair analysis may have psychological effects; therapeutic benefits, or other physiological effects, are unlikely for the following reasons:

(1) except for zinc, which appears to have some structural role, trace element levels in hair are not under homeostatic control—as are the levels of essential elements in many other tissues. Consequently a wide range of concentration is compatible with normal health.
(2) Trace element levels often vary widely among different hairs on the same head.
(3) These levels are greatly influenced by washing, by external contamination and by diet.

The claims made by commercial organisations offering hair analysis should be regarded with caution. It is true that anomalies associated with the intake of some elements in toxic amounts—or in people with serious nutritional disorders—are often reflected in the chemical composition of hair. Evidence obtained in this way can be useful in epidemiological studies—for example in the investigation of occupational hazards by comparing hair levels in exposed people and non-exposed controls. But for an individual patient the normal range of trace element concentration in hair is so wide that even conspicuous departures from expected values should be used only to support (or dispute) conclusions based on other evidence.

Some organisations offer to detect food allergies by analysis of hair. In a widely-publicised British trial, conducted in 1986, commercial laboratories were provided with samples taken from each of a number of people, some with known allergies and some with none. The outcome was not impressive. None of the laboratories agreed in their findings; allergies were found where none existed and, except in one subject, reported (incorrectly) as being allergic to almost everything, known allergies were not reported.

CHAPTER 2

All Flesh is Grass

Isaiah's assertion (though ignoring the contributions of air and water) summarises one stage in the transfer of trace elements from the earth's crust into the tissues of the body. The prophet might have added: 'All grass is rock', for the trace elements in plants are derived from the soil which is a product of the earth's crust.

Rocks are generally hard and impermeable and do not support much chemical or biological activity. Wind, rain and alternations of temperature start the breakdown into smaller fragments, thereby increasing the surface area available for chemical exchanges—of which the most immediately important is the formation of the soil solution by the incorporation of various mineral constituents.

Soil is the only source of trace elements for plants, from which virtually all food is obtained—either directly or after processing by animals. Some trace elements of great biological importance are present in soil at concentrations of 1 PPM or less. Others occur at concentrations which are very much greater, though not always adequate. Iron, which is a trace element in plants and animals, is a major constituent of soil—yet iron deficiency is common in plants and in people. The lesson conveyed by this paradox is that the availability of a trace element to a plant is more important than its actual concentration in the soil.

The availability of trace elements is influenced by many factors, of which the most obvious are soil acidity and drainage. Treatment with lime, to counteract acidity, facilitates the uptake of molybdenum and reduces the uptake of cobalt. Plants grown on wet (poorly drained) soil contain more cobalt, molybdenum and manganese than those grown on well-drained but otherwise similar soil. Clover takes up molybdenum more avidly than do grasses. For this reason it is sometimes necessary (for example, where soil is relatively rich in molybdenum, or has been treated with molybdenum to overcome a natural deficiency, or has been limed) to restrict the amount of clover in pastures, lest grazing animals suffer toxic effects from excessive intake of molybdenum.

The transformation of rocks into soil sometimes proceeds in a roundabout way. Artificial fertilisers, applied to increase the soil content of nitrogen, phosphorus or potassium, contain small amounts of several trace elements. Cadmium, copper, arsenic and selenium are sometimes present at much higher concentrations in the fertiliser than in the untreated soil. Plants grown on soil to which phosphate fertiliser has been applied over the years may contain undesirable amounts of cadmium.

Sewage sludge (the solid residue resulting from the treatment of waste water from domestic and industrial sources) contains useful amounts of nitrogen and phosphorus and is sometimes used as a substitute for manure or compost. Because of the discharge of industrial wastes into sewage systems, the dried sludge sold to farmers also contains several metals. If the origin is a town with considerable industrial activity, the sludge is likely to contain arsenic, cadmium, copper, mercury, lead and zinc at concentrations many times higher than in normal soil. These excesses are reflected only to a small extent in food plants grown on the treated soil. In Britain, the Department of the Environment has issued guidance intended to limit the contamination of food attributable to the use of sewage sludge.

THE SIEGE OF SHIPHAM

Contamination of soil by waste from mining and smelting may persist for centuries. In 1978 the village of Shipham in Somerset was invaded by geologists, chemists, doctors and journalists after cadmium, zinc and lead were found in unusually high concentrations in soil samples taken in locations where zinc had been mined during the 18th and 19th centuries. In vegetable gardens, to a depth of 15 cm, the soil contained cadmium at a mean concentration of 89 PPM, zinc at 8750 PPM and lead at 2356 PPM—all much above the normal level in Britain. Vegetables grown on these soils contained cadmium at a mean concentration of 0.28 PPM (winter) and 0.52 PPM (summer). Corresponding figures for lead were

1.44 PPM and 0.45 PPM. The villagers were advised not to eat home-grown vegetables and were invited to take part in an extensive series of health tests, including analytical studies on blood, urine and hair. No health effects were apparent and none has been found in subsequent enquiries.

Judgements on the hazards associated with trace elements in soil should be made cautiously. It is not always appreciated that the concentration of potentially toxic trace elements is substantially reduced at each stage in the sequence soil—plant—animal tissue—human foodstuff. This helpful discrimination mechanism is modified by animals which ingest appreciable amounts of soil while grazing. Soil commonly represents 1-10 percent of the dry matter eaten by cattle; the corresponding figure for sheep, which graze closer to the ground, may reach 30 percent or even more. Soil may be a useful source of essential elements, such as cobalt, copper, selenium and zinc. But in areas contaminated by previous or current industrial activity, ingested soil usually provides the greater part of an animal's intake of arsenic, copper and lead. The distinction between intake and uptake is important here. For the elements just mentioned, only a small proportion of the soil content is in soluble form; most is therefore unavailable to plants and is not taken up by the roots. For this reason the concentration of these elements in plants is usually much lower than in the soil on which they are grown. Information about the availability of trace elements in soil ingested by animals is meagre.

AFTER CHERNOBYL

An unpleasant modification of the soil-plant-food relationship discussed above came to light unexpectedly after the nuclear accident at Chernobyl in 1986, which released large amounts of radioactive fall-out, including caesium-134 (half life 2.1 years) and caesium-137 (half life 30 years). Stable caesium is relatively innocuous, but the passage of the radioactive isotopes into food chains is unwelcome. Since much of the fall-out comes down with rain, meat from animals produced in wet regions is particularly vulnerable. It was necessary to impose restrictions on the sale for slaughter of sheep from upland areas of Wales, Cumbria and Scotland. It was expected that these restrictions would be lifted in 1987, when the animals were grazing on freshly-grown herbage. As already mentioned, many metals are held in the soil in insoluble forms which are not taken up by plants.

Analysis of new grass in 1987 showed unexpectedly high concentrations of caesium, necessitating continued restrictions with serious economic consequences for hill sheep farmers. The reason for the continuing high concentration of caesium in the herbage was that grass in the affected areas grows on peat with a very thin soil cover. Consequently the fall-out (and the caesium excreted by the sheep) does not sink far enough into the ground, but remains in the root mat, from which the following year's growth emerges—still contaminated by radioactivity.

GEOCHEMICAL PROSPECTING

The chemical composition of soil, and of plants growing on it, is influenced to some extent by underlying deposits of minerals. For this reason, analysis of soil or plants is sometimes useful in prospecting for metals. This possibility was recognised as recently as the early 1930s, first by Russian geochemists and soon afterwards by workers in other countries. Relatively high concentrations of tin were found in plants growing in Cornwall, a long-established tin mining region. In Finland, the presence of nickel ore, at a depth of a few metres, was reflected in the unusually high nickel content of birch leaves.

Prospectors and commercial organisations using trace element analysis of vegetation or soil in searching for metal deposits do not always disclose their results or methods, but many successes have been recorded. Arsenic, which can be estimated with excellent sensitivity by activation analysis, is particularly relevant in this connection, because its appearance often indicates the proximity of more valuable elements, such as gold and silver. Young Douglas fir trees are useful indicators; the normal arsenic content in the ash of first or second year growth is less than 10 PPM, but concentrations of thousands of PPM may be found in association with precious metal deposits. Gold and silver ores often contain unusually large concentrations of lead, tellurium and mercury, as well as arsenic. All of these elements can be detected in soil and vegetation. Where uranium and selenium occur together (as in Colorado), the mere presence of some species of astragalus (locoweed, p 62) is significant, as these plants need great amounts of selenium.

Trace element analysis led to a major discovery of gold at Cortez, Nevada, in 1966. Investigations elsewhere in Nevada showed in 1964 that unusual amounts of antimony, tungsten, arsenic and mercury, found in gold ore, were present also in soil from the neighbourhood of a mine. Then these metals were detected in soil at Cortez, opencast exploration for gold revealed a deposit of three million tons of ore, with an average content of 0.3 ounce of gold per ton.

WATER AND HEALTH

Water is best, proclaimed Pindar, more than 24 centuries ago. When man's environment was dominated by dirt and disease, water—whether applied externally or internally—was a natural symbol of purity. During the 19th century, when the practice of medicine appeared to have little effect on the health of the people, hydropathy developed into a complete alternative system. Some authorities advocated cold baths, wet sheets, vigorous exercise and 40 glassfuls to drink every day. Others favoured a different regime. In the centuries before public water supplies had been effectively separated from sewage, access to a clean and colourless source was welcome. But the belief that medicines with a pleasant taste (or no taste at all) were lacking in therapeutic virtue was not easily dispelled. All over Europe, invalids flocked to the spas to take copious draughts of water with a foul taste and a bad smell. The spas, like the health farms of today, also helped to ease the feelings of guilt among clients given to over-indulgence in food and drink. The benefits claimed for spa treatment (and, to be fair, often perceived by the customers) were largely attributable to the high fees and frugal diet.

During the present century the provision of wholesome water became a matter of municipal pride. People living in soft water areas, where the tap delivered virtually pure H_2O, thought themselves fortunate, for their tea tasted better, their soap lasted longer and the insides of their kettles and water pipes remained free from the deposits of carbonates of calcium and magnesium which accumulated in hard water areas.

But now we are not so sure. The first warning came in 1957 from Dr Jun Kobayashi, who found that there was a positive relationship between the incidence of fatal strokes in different parts of Japan and the acidity of the local water supplies. (A stroke is the result of cerebral hemorrhage, usually associated with high blood pressure). Since then there have been many other studies on the relationship between water quality and cardiovascular disease—a term which embraces all diseases of the heart and circulation, including coronary thrombosis. The relationship is important because cardiovascular disease accounts for about half of all deaths in Britain.

HARD OR SOFT WATER?

The customary index of water quality is the hardness—a property which is easier to describe than to define. All water supplies are maintained by rain or snow. In Scotland, Wales and northern and western England, most of the rocks below the surface are igneous and impervious. Water supplies are therefore obtained from natural or man-made reservoirs fed by streams originating on high ground. Rainwater, after appropriate treatment in reservoirs, is almost entirely free from dissolved matter and for this reason is very soft in quality.

In southern, eastern and midland regions of England, where the underlying rocks are mostly of sedimentary origin, including limestone, chalk and sandstone, rainwater percolates deep into the ground and is recovered from wells or boreholes. Supplies obtained in this way (or from rivers which flow over sedimentary rocks) contain appreciable amounts of the sulphates and carbonates of calcium and magnesium, which produce the characteristic properties of hard water.

Calcium carbonate is the principal constituent of interest in hard water. Hardness is expressed in milligrams per litre (equivalent to PPM) of calcium carbonate—in practice, the concentration of calcium carbonate which would produce the same effect, in laboratory tests of hardness, as the mixture of calcium and magnesium salts present in the water. Colwyn Bay, at 10 mg/l, has the softest water in Britain and Hartlepool, at 528 mg/l, has the hardest.

In considering this problem, it is useful first to examine the evidence and then to speculate on possible explanations. The evidence is of two kinds. There is a great deal of information on geographical variation in death rates from cardiovascular disease in Britain. In general, the death rate (adjusted to allow for differences in age distribution) is higher by about 30 percent in towns with soft water than in those with hard water. It is of course necessary to allow for the influence of many other relevant factors—climatic, economic, environmental and social. There is, for example, a positive correlation between cardiovascular death rate and several other factors, including rainfall, percentage of unskilled or manual workers in the community and percentage unemployment. There is a negative correlation with average temperature and number of cars per household.

When adjustments are made for these factors, the death rate for cardiovascular disease is about 10 percent higher in towns with soft water than in those with hard water. There are considerable variations among individual towns, both in soft water and hard water areas. People living in two small Scottish towns (Hamilton and Airdrie) are twice as likely to die of cardiovascular disease between the ages of 35 and 74 as their contemporaries in towns on the outskirts of London.

The disturbing effect of factors other than water hardness can be avoided when statistics are gathered from towns which have changed their water supply, from soft to hard or from hard to soft. A careful study of this kind was reported by the Water Research Centre in 1981. Between 1961 and 1971, four towns (St Helens, Sunderland, South Shields and Wigan) reduced the hardness of their supplies by more than 25 percent. Ten towns increased hardness by more than 25 percent. A group of towns where no change occurred served as controls. Statistical analysis showed, among men (and less markedly among women) a significant trend in the direction of decreasing cardiovascular mortality with increasing water hardness.

It is unlikely that investigations of this kind can be improved much further. Increase of water hardness beyond 170 mg/l appears to have little or no effect on mortality from cardiovascular disease. Convincing effects might be expected from reduction of hardness to the very low levels of 10-20 mg/l prevailing in a number of Scottish and Welsh towns—but this is an experiment that no water authority wants to make, in view of the adverse health effects associated with very soft water. In any event, a current EEC directive requires that the hardness of public water supplies must not be reduced artificially below 150 mg/l. The lack of consensus in this matter is illustrated in advice given by the US National Research Council in 1977 and by the World Health Organisation in 1984. Both concluded that there is not enough evidence on health grounds to justify national policies to alter either the hardness or the softness of water supplies.

Do-it-yourself improvement of water quality has become popular in hard water areas, using devices which are easily fitted to domestic plumbing systems. The principle involved is ion exchange. Water is passed through a column of synthetic resin which removes calcium and magnesium, replacing these elements by sodium. The water is certainly softened, but its sodium content is considerably enhanced. Many authorities believe that increase in the intake of sodium (which is already abundant in most countries) adds to the risk of increasing blood pressure. It is therefore advisable to have a separate outlet, bypassing the softener, for drinking and cooking.

It is generally believed in the water supply industry that hardening of soft water supplies, on a national scale, would not be justified, partly because of the substantial capital and running costs but also because the benefit to be expected is not very great. Changing a water supply from very soft to hard reduces the chance of death from cardiovascular disease by about 10 percent—but a heavy smoker is twice as likely to die from cardiovascular disease as a non-smoker.

WATER AND HEART DISEASE

A reduction of 10 percent may not appear impressive—but it is 10 percent of a very large number. If cardiovascular mortality in Britain could be reduced everywhere to the level prevailing in hard water areas, the annual number of premature deaths prevented would be about the same as the annual number of deaths from road accidents. It is therefore worth while to examine in more detail the relationship of water hardness to cardiovascular disease.

This is a difficult problem. Most substances harmful to health exert their effects when present in unduly large amounts. But soft water is a very pure substance, containing virtually nothing but H_2O. Perhaps some of the elements present in hard water are essential for health? This is most unlikely, since the elements concerned are all present to a much greater extent in the diet. Water with a hardness of 170 mg/l (above which there is very little effect on cardiovascular mortality) will augment the normal dietary intake of calcium and magnesium by no more than 10 or 15 percent, which is well within the normal variation among individuals. Perhaps elements more abundant in soft water are harmful? Soft water is usually slightly acidic and therefore dissolves lead from old-fashioned plumbing systems. Most of the lead in the body is stored in bone, where it accumulates with increasing age. In 1969, before the Glasgow water supply was artifically hardened (p 105) *post-mortem* measurements of lead in dried rib bones (from two groups matched for age and sex) found mean levels of 83 PPM in Glaswegians and 44 PPM in Londoners. But exposure to lead is not recognised as a factor in cardiovascular disease.

Perhaps water hardness is irrelevant. In Japan, where virtually all water supplies are soft, the significant property appears to be acidity—more exactly, the ratio of sulphate to carbonate. In Britain also, this ratio is related to death rates from stroke, but less markedly than water hardness. It might be thought significant that the hard water towns

SCRAPMETAL
DEALER
COPPER LEAD BRASS ETC

SPARE
RIBS

FLEMING

Most of the lead in the body is stored in bone.

in Britain are mostly in the prosperous areas of southern England, whereas the soft water towns are mostly in the less affluent areas of Scotland and northern England. Speculation on this basis is not helpful, since water hardness has little influence on mortality from non-cardiovascular diseases. The relation of water to heart disease remains a tantalising problem.

A MYSTERY RESOLVED

Trace element analysis has recently provided the solution to another mystery related to water quality. This problem arose out of ingenious—and apparently successful—efforts to imitate the body's water purification system. Every drop of water that we drink is passed through the kidneys more than 100 times before being discarded in the urine. At each passage, unwanted materials are filtered out, with an economy of space and energy consumption that chemical engineers admire but cannot emulate. When the kidneys cannot cope, it is necessary to provide an external system, to remove toxic substances from the blood and to maintain essential materials at the appropriate concentration by the process of haemodialysis.

This is done in the artificial kidney, a machine thousands of times larger—and much less efficient—than the organ that it simulates. Blood released from an artery passes through a thin-walled tube or sandwich of a suitable plastic material. On the other side of the membrane is a solution of various salts. Urea and other potentially toxic materials pass through the membrane and are removed from the blood, to the patient's great advantage; the consequences of poisoning or drug overdose can often be forestalled in the same way. On the other hand, essential materials can be added to the blood by dissolving them at suitable concentrations in the dialysing fluid. Haemodialysis can be carried out in hospital or, with the provision of appropriate equipment, in the home.

The purity of the water used to make the dialysing fluid is important because of the quantities involved. A healthy person drinks between one and two litres of water each day. In the artificial kidney, the patient's blood exchanges material with 50–100 litres of water per day. For reasons of cost and convenience, tap water is generally used to make up the dialysis fluid. During the 1970s, evidence accumulated to show that this practice is not always safe. It was observed, first in the United States and soon afterwards in other countries, that some patients treated by dialysis suffered brain damage, which was usually fatal within a few months. The disease was given the neutral name of dialysis encephalopathy, but its cause was not uncovered for several years.

Aluminium was suspected, at first for the wrong reason. Patients undergoing haemodialysis are liable to become dangerously overloaded with phosphorus, because their defective kidneys are unable to remove enough of the element from the blood. This problem is dealt with by dietary restrictions and by giving aluminium hydroxide or carbonate. These salts combine with dietary phosphates to form insoluble compounds which are not absorbed from the intestine and are therefore safely eliminated from the body. Unfortunately, aluminium hydroxide and carbonate are themselves absorbed to a small extent from the intestine. In a healthy patient this is not dangerous, because aluminium is excreted by the kidney—but abnormally high concentrations of aluminium were found in the brain and other tissues of victims of dialysis encephalopathy. By 1980 it was established that water was the significant source of the trouble and that dialysis encephalopathy could be almost completely avoided by the removal of aluminium from water used to make dialysis fluid.

An interesting study reported in 1978 supported this finding. Among patients living in the west of Scotland who had undergone haemodialysis at home betwen 1972 and 1977, there were 13 deaths from dialysis encephalopathy. Four of the victims lived in Dumbartonshire, four in Dumfries-shire and five in Lanarkshire—but there were no cases among the 40 patients living in Glasgow. Investigation revealed that aluminium sulphate was added to public supplies (to clarify the water by removing organic matter) in the areas where deaths had occurred, but not in Glasgow. Atomic absorption measurements showed that the mean concentration of aluminium in water in the homes of the 13 victims was 0.4 PPM; in the 40 Glasgow homes the corresponding figure was below the detection limit of 0.04 PPM.

THE POLLUTION PARADOX

Concern over pollution is greater today than at any time in the past, although both the internal and external environments are cleaner than for many centuries. This paradox reflects changes in the chemical nature of pollution and, more significantly, changes in public perception of the external environment. The literature of Greece, Rome and other ancient civilisations contains many references to the pastoral ideal, embodying the belief that a simple life, without technology, commerce or industry, was man's natural state, ensuring peace, health and happiness, and that it existed in a Golden Age from which society had deteriorated. Thomas Jefferson thought that the instinctive yearning for the simple rural life might be fulfilled on a large scale in the United States. 'Those who labour in the earth' he wrote, 'are the chosen people of God . . . let us never wish to see our citizens occupied at a workbench . . . let our workshops remain in Europe'.

He could not hold back the tide of industrial activity, or of the pollution that grew with it. But most people were indifferent, or even hostile, to the preservation of the natural environment. When Dr Samuel Johnson made his celebrated journey through the highlands of Scotland in 1773, he complained about the wilderness of desolation and the uncultivated ruggedness. When Madame de Stael went out from her chateau in Switzerland, she ordered the carriage blinds to be drawn, because she found the sight of the mountains so upsetting.

Pollution was unrestrained during the 19th century but few people objected. Today the situation is very different. Pollution has decreased but the public expectation of environmental quality—related both to health and to amenity—has risen greatly. Every large-scale advance in technology, whether for energy production, manufacturing industry, communication or agriculture, is subject to close scrutiny—and sometimes to vigorous objection—because of its environmental impact. Changes in public perception of environmental management are accompanied (and to some extent provoked) by changes in the nature of pollution. Until the beginning of the 20th century, the material environment was largely composed of materials (such as wood, animal and vegetable fibres, metals and other minerals) used in their natural state or processed in unsophisticated ways. When such materials are discarded they decay by natural means into products which may eventually be recycled and which, in any event, do little harm.

Today we are adding to the environment great amounts of substances—such as plastics, drugs and other chemicals—which were not known a century ago, which do not decay after a reasonable time into harmless remains, and whose effects on the health of plants, animals and man are imperfectly understood. In these ways, as well as by long-established manufacturing processes, much contemporary industrial activity releases potentially toxic substances into the environment. The resulting health hazards are sometimes exaggerated (in contrast to the hazards associated with alcohol and tobacco, which are generally accepted with equanimity or even embraced with enthusiasm) but much uncertainty remains. In the tasks of identifying pollutants, finding their sources and estimating their significance, trace element analysis has achieved some interesting successes—for example, in the investigation of accidental leakage from oil tankers.

OIL SPILLAGE

The sea was calm and the sun was shining on Saturday morning, 18 March, 1957 as the Torrey Canyon, one of the world's largest ships, passed Lands End on her way to Milford Haven with a cargo of crude oil from Kuwait. Ignoring the warnings on the navigation charts—and the flare and rocket signals given in ample time by the crew of a lightship—the tanker ran onto a reef, holed 6 of her 18 storage tanks and discharged 95 000 tons of oil into the sea. This disaster, which did great damage to fish and bird life, and contaminated beaches as far away as Normandy, focussed public attention on the increasing hazard of pollution at sea.

Accidental or deliberate spillage from ships represents only a small proportion of the total quantity of petroleum products released into the sea. Most of the discharges come from normal shipping operations, offshore wells, refineries, and oil waste from industry and motor vehicles which eventually reaches the oceans. These discharges are widely distributed in space and time, but large accidental spillages are localised and produce more obvious effects.

TRACING THE POLLUTER

The economic attractions of supertanker operations have been eroded by heavy anti-pollution insurance premiums. Since the ship's operator (or insurer) may be liable for damages, it is important to have some way of identifying the culprit after a spill. This is not as easy as it appears, for the oil slick (and the ship) can travel a long way before the pollution is observed. Analysis of a small sample from an oil slick can give useful information as to its origin. Oil is a very suitable material for activation analysis. Its major constituents, hydrogen and carbon, do not become radioactive to a significant extent, but many of the commonly-occurring trace elements are easily detected and measured.

There is a wide variation in trace element content among oils from different sources. Analyses of crude oil samples made during a major study in California in 1969 gave the following results:

Origin	Vanadium, PPM
California	53.3
Venezuela	177
Louisiana	1.0

As similarly wide variations were found for other elements, it was established that measurement of trace element concentrations in a sample of oil slick would give a reliable indication of its origin.

Trace element analysis was used to investigate pollution which occurred in Nova Scotia when the tanker Arrow was wrecked in 1970, with the loss of several thousand tons of fuel oil. A few weeks later, heavy oil pollution was reported from Sable Island, 100 miles from the wreck. Samples of oil from the wreck and from the polluted beach were analysed with the following results:

	Oil from wreck	Oil from polluted beach
vanadium, PPM	332	352
nickel	39	41
copper	2	2
vanadium/nickel	8.5	8.6

The ratio of vanadium to nickel in the samples was calculated for the following reason. Over a period of weeks, some of the volatile components of the oil evaporate. The trace metals are, however, usually present in chemical forms which are not volatile. The relative concentrations of these elements are therefore not changed by evaporation elsewhere in the sample. The closeness of the vanadium/nickel ratios indicates that the two samples had a common origin, giving support to the inference that oil from the wrecked tanker contaminated the distant beach.

ATMOSPHERIC POLLUTION

Atmospheric pollution, which travels over even longer distances than oil slicks, has stimulated considerable scientific study during the past 20 years. The initial impetus came from concern over the more obvious effects of pollution, including smoke, fog, dirt and respiratory disease. These problems have been eased by smoke control legislation, restriction of industrial emissions and the change from domestic coal fires to central heating. But there is increasing recognition of the lack of an adequate scientific basis for the controls (now being implemented at great cost) on emissions from power stations, factories and motor vehicles.

Trace element analysis contributes usefully to the modest research effort now concerned with the large-scale movement of atmospheric pollution. The atmosphere contains (from natural processes greatly augmented by industrial and domestic activity) large amounts of aerosols. An aerosol is a suspension of minute solid and liquid particles, mostly no more than a thousandth of a millimetre in diameter. An aerosol mass, like an oil slick, usually has a chemical composition which is typical of its original source, and therefore (in principle and to a large extent in practice) allows its movements to be plotted. Sampling of aerosols is done by drawing air through a filter or other collector. In an urban environment, aerosol lead comes almost entirely from motor vehicle exhausts; its amount depends on progress in removing lead from petrol. Vanadium comes almost entirely from oil burning, and zinc from refuse incinerators. Aluminium and iron are released in coal burning, but are found also in aerosol particles from soil; a distinction can be made, because arsenic is more abundant in coal than in soil, while manganese is more abundant in soil.

RECEPTOR MODELLING

The movement of air pollution can be studied by the technique of receptor modelling. This is conducted in two stages, involving the source area (where pollution originates) and the reception point, where samples of aerosol particles are collected for analysis. In the first stage, information is obtained on the trace element composition of the aerosol material released from each of the major relevant categories of activity (for example, smelting, refining and power generation) in the source area—which may be an industrial complex, a city or even a whole country. Typical trace element profiles for motor vehicle exhaust and for domestic coal and oil burning are also included. It is feasible to use as many as 20 elements for this purpose, though usually fewer will suffice.

If the trace element composition of the aerosol mass does not change as it travels from the source area to the reception area, and if each activity category releases aerosol with a different trace element concentration pattern, and if the number of elements measured is at least equal to the number of pollution producer categories, it is theoretically possible (by measuring the concentration of each trace element in the sample taken at the reception area and solving a set of simultaneous equations) to calculate the contribution of each category of pollution producer to the aerosol mass.

In practice, there are several reasons why a completely accurate solution is not achieved—for example, heavier particles may settle out of the aerosol as it travels and other particles may be gathered as a result of industrial activity along the way. Studies of the kind described here do nevertheless give useful information on the relative contributions of different types of source.

NOT-SO-RARE EARTHS

Receptor modelling, using rare earth elements as tracers, is particularly useful in relation to emissions from oil-burning power plants and oil refineries. The rare earths, which are very easily detected and measured by activation analysis, are present in the zeolite catalysts now used to an increasing extent in oil refineries. Some of the catalyst material escapes in waste gases and some appears in the products; oil refineries in the United States release more than 7000 tons of rare earth elements into the environment every year. The trace element profiles of emissions from different sources are prepared by calculating the relative concentrations of lanthanum and other rare earth elements in the emitted aerosol. Table 2.1 illustrates the ease with which emissions from power stations and refineries can be distinguished. The rare earths are particularly useful for receptor modelling over long distances; because of the close chemical similarity among these elements, their relative proportions are not likely to be disturbed by processes which might alter the contributions of other elements to the trace profile.

During recent decades, legislation has affected considerable improvements in air quality. In many industrial countries smog has disappeared and fog has become a rare occurrence. In New York City the concentration of sulphur dioxide in the air is only a tenth of what it was 25 years ago and the amount of dust in the streets of Manhattan has fallen even more sharply. The long-term prospect is not so clear. Pollution of the atmosphere and stratosphere produces a variety of effects which are not yet sufficiently understood. Accumulation of aerosols in the stratosphere increases the proportion of solar radiation reflected back

Table 2.1 Concentration ratios of lanthanum to other rare earth elements and to vanadium in particles emitted from various sources of atmospheric pollution.

	Coal fired power plant	Oil fired power plant	Oil refinery
La/Ce	0.51	1.8	1.25
La/Nd	1.6	2.9	1.83
La/Sm	5.2	28	20
La/Yb	18	135	950
La/Lu	80	1000	5400
La/V	0.3	0.045	20

into space and therefore tends to lower the global temperature. In the stratosphere, the temperature may increase, because aerosols absorb heat from the warmer atmosphere and surface below more effectively than they lose it into space.

THE OZONE LAYER

It has been suggested that the ozone layer in the stratosphere (roughly 6-30 miles above the Earth), which absorbs biologically harmful short wavelength ultra-violet radiation from the Sun, is vulnerable to erosion by chlorofluorocarbons, composed of carbon, chlorine and fluorine; the commonest have the formulae $CClF_3$ and CCl_2F_2. These substances have many uses. They are present in refrigerators and air conditioners, in rigid and flexible polyurethane and polystyrene foam, in dry cleaning fluids and as propellants in spray cans.

From all of these applications they are released into the environment. They are present in the lower atmosphere at concentrations much less than one part per billion. When they eventually make their way into the stratosphere they gradually decompose, releasing chlorine which catalyses the breakdown of ozone into the oxygen.

It was reported in 1987 that the ozone layer above the south polar region has already been punctured by a hole roughly the size of the United States. Loss of the protection provided by the ozone layer would affect plant growth (and therefore food production) and would increase the incidence of skin cancer. In 1987 the major industrial countries agreed that the production of chlorofluorocarbons should be reduced by about 35 percent before the end of the century. The magnitude of the hazard to the ozone layer is, however, still a matter of debate. Considerable amounts of active chlorine and fluorine compounds are released from volcanoes, without apparent effect on the ozone layer. Dr Jim Lovelock has pointed out that gases and vapours produced in the Krakatoa eruption of 1883 could have depleted the ozone layer by 30 percent.

The environment is vulnerable in other ways, because the inevitable growth in consumption of fossil fuels, and in industrial activity generally, increases the contamination of the atmosphere by carbon dioxide and so enhances the greenhouse effect. Radiation from the Sun, where the surface temperature is about 6000°, is of short wavelengths and passes, without much loss, through the air to heat the Earth. Heat re-radiated from the ground is of much longer wavelength and is absorbed by carbon dioxide in the air, with a consequent increase in global temperature.

The importance of the contribution of aerosols to the greenhouse effect is uncertain. Some evidence has come from spacecraft mission to other planets. Venus is covered by a thick aerosol layer, rich in sulphuric acid; a much thinner layer of similar chemical composition, but smaller particle size, occurs in the Earth's atmosphere. It is believed that the Venusian aerosol layer, because of its larger particle size, heats the plant by greenhouse effect. There is, however, much more to be learnt by the study of aerosols on Venus and on Mars.

POLLUTION IN THE ARCTIC

Receptor modelling might help in studying air pollution in the Arctic. This story began in 1957, when Murray Mitchell first reported atmospheric turbidity (often referred to as haze) extending from Northern Alaska to the polar regions during winter months. Hazes of natural occurrence are not unusual and Mitchell's discovery did not alter the long-established belief that the Arctic and Antarctic atmospheres were singularly free from pollution. Since 1972 it has gradually become apparent that the atmosphere of the whole Arctic region is polluted during the winter. The haze consists of bands of aerosol, which may be thousands of kilometres wide and between one and three kilometres deep. Research so far has used measurements of a few trace elements (aluminium, vanadium, manganese, sodium and barium) in aerosol samples, taken from aircraft or at ground level, at Barrow (latitude 71°

North) in Alaska. Measurements have also been made of oxides of sulphur and of lead-210, a naturally-occurring radioactive isotope which is serviceable as a tracer in studying large-scale air movements.

The sources of the Arctic pollution are still being debated. Much of it appears to come from the USSR and from Europe; no more than 10 percent is attributed to North American sources. Many other interesting problems remain to be solved. The presence of bromine in the Arctic air at a concentration which, between February and May, is higher than in any non-urban region in the world, is a mystery. It has also been found that the air contains measurable amounts of several gaseous pollutants, including methyl chloroform and Freon. The term 'Arctic haze' should, it is suggested, connote the whole polluted air mass, and not just the aerosol content.

Kenneth Kahn, a pioneer in Arctic air studies, wrote in 1985:

> 'the Antarctic is clean and pure, pristine and noble . . .. By contrast, the Arctic is indeterminate, broken up and dirty. The Antarctic represents the pristine character that civilisation has lost and dreams of somehow regaining. The Arctic is what we as an industrialised hemisphere have really become; it mirrors our societal faults'.

. . . Which brings us back to the thoughts expressed on page 30.

THE END OF THE DINOSAURS

Atmospheric pollution of an unusual kind has recently been invoked to explain the disappearance of the dinosaurs—a mystery which has long been a topic of vigorous discussion among biologists and palaeontologists.

The latest contribution to the debate illustrates the enthusiasm with which Nobel prize winners, after reaching the pinnacle of scientific achievement, sometimes strike out in new directions. Linus Pauling (chemistry, 1954) became active in politics (winning the peace prize in 1962) and advocated the study of vitamin C as a cure for cancer. Brian Josephson (physics, 1973) investigated the relevance of Eastern mysticism to contemporary physics, Martin Ryle (physics, 1974) espoused the cause of wind power as an alternative to nuclear generation and William Shockley (physics, 1956) promoted controversial views on intelligence and race. Luis Alvarez (physics, 1968) applied his insight and ingenuity to the attack on an old and fascinating scientific question; what happened to the dinosaurs—or, rather, how did it happen?

The dinosaurs were the masters of the world for 150 million years. Not all of them were the clumsy creatures of the school room and the comic strips; archaeopteryx was no larger than a crow. The dinosaurs were a highly successful group of animals, well adapted anatomically and physiologically for survival and development. Yet they disappeared some 65 million years ago, along with about half of all the living world. This catastrophe, the latest in a series of extinctions revealed in the fossil record, which goes back for nearly 600 million years, is used to define the boundary between the Cretaceous and the Tertiary periods of geological history.

Discussion of the Cretaceous—Tertiary (C–T) extinction has generated much controversy and a variety of explanations, including poisonous plants, reversal of the Earth's magnetism, flooding from an Arctic lake and radiation from a nearby supernova. In the most convincing of the theories now on offer, Alvarez and his team—which includes his son (a professor of geology) and two nuclear chemists, suggests that the key to the mystery lies in the seabed sediments laid down at the time of the C–T extinction. It might be thought that this material would by now be far under the sea floor, buried by sediments deposited during the past 65 million years. Some of it is there, but much of it is above ground. Most of the mountains contain fossilised remains of tiny creatures which once lived in the sea. Organic remains have, of course, long since decayed, but the calcium carbonate shells of the diatoms, *foraminifera* and such like creatures have survived unchanged, along with other material released by erosion of the Earth's crust and washed into the seas.

Mountain ranges all over the globe have been formed by upheavals of the ocean floor. It is therefore possible to examine (and, by various techniques, to date) the sediments laid down in times past. Limestone rocks normally contain clay, washed from the land by rivers into the sea, mixed with calcium carbonate from the shells of *foraminifera*. Clues to the C–T extinction are hidden in a layer of clay, about a centimetre thick, which is to be found in limestone rocks in many parts of the world; more than 70 sites have so far been identified. Shells of *foraminifera* are absent from this boundary layer, though they can be seen in abundance above and below it. Confronted by this curious observation, palaeontologists had to presume that the deposition of calcium carbonate on the ocean floor had, for some reason, stopped for a while 65 million years ago.

In 1979 Alvarez speculated that an explanation for the extinction of the *foraminifera* might be found by measuring the concentration of iridium in the boundary layer. This was not a shot in the dark, because iridium had for many years been an element of interest to geochemists. This interest arises, oddly enough, from the fact that there is hardly any iridium in the earth's crust. When the primitive Earth was cooling from a molten state, iridium and other metals of the platinum group (including osmium, palladium, rhodium and ruthenium) were swept into the core, along with iron. Consequently the average concentration of iridium in the Earth's crust (less than 0.1 PPB and too small to be measured accurately before the arrival of activation analysis) is very much less than in the Universe, as estimated by the analysis of stony meteorites, which contain about 0.5 PPM of iridium. The concentration of the platinum group metals in sedimentary rocks is—though still very small—greater than in other components of the crust. It was suggested in 1952 that this anomaly was attributable to the deposition of meteoritic dust, released by the intense heating of meteorites passing through the Earth's atmosphere.

Iridium can be detected and measured with great sensitivity by neutron activation analysis. Alvarez thought that measurement of the iridium concentration in the clay layer corresponding to the C–T boundary would show whether the extinction had been triggered by the impact of an asteroid—one of the miniature planets which occasionally stray into the Earth's orbit. An event of this kind would create a fireball and release a great quantity of dust—enough to obscure the Sun for several months. The temperature would fall to about – 18°C for many months. The radiant energy of the fireball would produce oxides of nitrogen in the atmosphere and the resulting acid rain, falling into the oceans as virtually pure nitric acid, would wipe out the *foraminifera* by dissolving their shells. Photosynthesis would stop and all marine and land plants would die. Animals, having no food, would also die. The impact of the asteroid would increase the concentration of iridium near the surface of the Earth, since the element is so much more abundant elsewhere in the Universe.

Alvarez first analysed samples taken from the boundary layer, and from the limestone above and below it, at Gubbio in Northern Italy. The iridium concentration above and below the boundary layer was about 0.3 PPB. In the boundary layer it rose to 9.1 PPB. To find whether this was merely a local effect, samples were obtained from the C–T boundary layer at a site in Denmark. There the iridium concentration in the boundary layer had a peak value of 41.6 PPB. Other analyses showed that the relative proportions of the various platinum group elements in the boundary layer were very similar to those found in stony meteorites.

Investigations at other sites, in many parts of the world have given results consistent with those found in Italy and Denmark. Alvarez estimates that the iridium content of the world-wide boundary layer is about half a million tons. To deliver this amount an asteroid containing 0.5 PPM of iridium would have to be about 10 kilometres in diameter. The energy released by the impact of such a missile, travelling at about 25 kilometres per second, would be equivalent to the explosion of 100 million megatons of TNT—some thousands of times greater than the total stock of nuclear weapons in the world today.

Palaeontologists are notoriously argumentative. It is not surprising that some of them are less than delighted at the success of an intruder from another discipline who has, with great panache, played them at their own game by applying sound scientific ideas and techniques to the attack on a difficult problem. The Alvarez theory is now widely respected, though a few questions remain to be considered. Why, for example, did the dinosaurs disappear while other animals survived? Where is the asteroid's crater, which must have been at least 100 kilolmetres in diameter? One possibility is that it has disappeared from view; some 20 percent of the Earth's crust 65 million years ago now lies under the continents. There is also a possibility that the crater lies on the ocean floor in a region which has not yet been surveyed with sufficient accuracy. The asteroid theory has been subjected to other criticisms, many of which are not very substantial—but the arguments will continue.

BURIED TREASURE

Some of the pollutants generated by human activity—and by natural disasters—find their way to the sea bed; but accumulations of useful trace elements are also found there.

The voyage of HMS Challenger, which lasted from 1872 till 1876 and took the 2300-ton frigate around the world, was one of the greatest scientific enterprises of all time. The five scientists who led the expedition made a massive contribution to biological knowledge and founded the modern science of oceanography. The expedition's major achievement was the discovery of the continental shelves, from which oil and gas are now being recovered in the North Sea and elsewhere. (One of the findings of the Challenger expedition was put to immediate use. Currents of cold water, far below the surface, were regularly sampled to cool the scientists' champagne!)

The Challenger team also discovered a vast accumulation of trace elements. The ocean floor in many places is covered by nodules, up to 30 cm in diameter, containing manganese at concentrations up to 40 percent, iron (up to 20 percent), nickel and cobalt (each about 1 percent) and smaller amount of cobalt and zinc. The nodules usually consist of layers of material surrounding a nucleus—often a shark's tooth. The origin of the nodules is still debatable. It is believed that some of the material represents the remains of plankton and other organisms living near the surface, which have the ability to concentrate metals from sea water. The size of the deposits is enormous. The Pacific Ocean alone is believed to contain some 1600 billion tons of nodules. One region to the west of Mexico is thought to offer up to 60 sites, each containing about 70 million tons.

The economic importance of the nodules is often exaggerated by politicians. The manganese and iron components are not of great importance, since these metals are abundant near the surface of the earth. Commercial exploitation is likely to be confined to the parts of the ocean floor containing nodules with more than 1 percent of nickel and of copper. There are, however, serious legal and political issues to be resolved. Many underdeveloped nations have been led to believe that a share in the riches of the seabed will solve their economic problems—but the capital investment needed to establish a nodule recovery project is of the order of a billion dollars; many such projects would be needed. If the necessary international agreement can be reached (probably under United Nations auspices) and if the required financial support is available, wealth should begin to flow from the ocean floor by the end of the century.

CHAPTER 3

The Inner Man

'I am a little world made cunningly of Elements'
(Donne, *Holy Sonnets, V*)

Living matter is made by the transient rearrangement from the crust, oceans and atmosphere. In this process some elements, such as carbon, oxygen, sulphur and calcium, are selectively concentrated at the expense of others, such as hydrogen, helium and silicon. Table 3.1 shows the approximate abundances of the ten commonest elements in each of several environments—Universe, continental crust of the Earth (excluding the oceans), sea water, kale (a green plant) and the human body.

In the geological and social enviroments, trace elements are those present in very small amounts. In the biological environment a further consideration is important, because it appears that some trace elements have an essential role in living systems, while the presence of others is merely fortuitous.

ESSENTIAL TRACE ELEMENTS

A trace element may confidently be regarded as essential if all of the following conditions are satisfied:

(1) Intake rate below a certain level leads to death or to impairment of some vital function, such as growth or reproduction.

Table 3.1 Percentage of total number of atoms. (NB This is not the same as the percentage by weight.)

Universe		Continental crust of the Earth	
hydrogen	92.8	oxygen	60.4
helium	7.1	silicon	21.4
oxygen	0.05	aluminium	6.3
neon	0.02	hydrogen	2.9
nitrogen	0.018	calcium	2.1
carbon	0.008	iron	1.8
silicon	0.002	sodium	2.1
magnesium	0.002	potassium	1.1
potassium	0.0015	magnesium	1.1
sulphur	0.0009	titanium	0.2

Sea water		Human body		Kale	
hydrogen	66.2	hydrogen	62.9	hydrogen	62.4
oxygen	33.1	oxygen	25.5	oxygen	30.6
chlorine	0.33	carbon	9.4	carbon	6.0
sodium	0.28	nitrogen	1.35	nitrogen	0.5
magnesium	0.05	calcium	0.23	calcium	0.16
sulphur	0.017	phosphorus	0.22	potassium	0.1
calcium	0.006	sulphur	0.05	sulphur	0.08
potassium	0.006	potassium	0.04	phosphorus	0.02
carbon	0.0015	chlorine	0.03	sodium	0.02
bromine	0.0005	sodium	0.03	chlorine	0.02

(2) The adverse (but non-fatal) effects of deficiency can be avoided or corrected by increasing the intake of the element—but not in any other way.

(3) The element has an identifiable role—which can be filled by no other element—in some physiological or biochemical process necessary for the maintenance of normal function.

It is difficult to obtain convincing evidence on all three of these points—even in laboratory experiments on plants and animals, where the intake of nutrients can be strictly controlled. Allocation of essential status to an element often starts with a chance observation, followed by empirical tests to confirm that conditions (1) and (2) are fulfilled. Conclusive evidence for condition (3) may be long delayed.

Progress in trace element research has been decisively influenced by advances in analytical chemistry. At the end of the 19th century, only two elements (iodine and iron) were known to be essential for human health. By 1935, only four (copper, manganese, zinc and cobalt) had been added. But progress has been more rapid during the past 40 years (table 3.2), largely because the revolution in analytical chemistry has greatly enhanced the experimenter's capability to measure trace elements in the extremely small quantities present in plant and animal tissues and in food.

Table 3.2 Trace elements: date of recognition of essential role.

iron	17th century
iodine	19th century
copper	1928
manganese	1931
zinc	1934
cobalt	1935
molybdenum	1953
selenium	1957
chromium	1959
tin	1970
vanadium	1971
fluorine	1971
silicon	1972
nickel	1974
arsenic	1975
cadmium	1977†
lead	1977†

† Evidence incomplete

The main reason for the extraordinary potency of the essential trace elements is that most of their work is done as components of enzymes or hormones. Enzymes are catalysts; they facilitate and accelerate chemical reactions (for example in the body), without being consumed or altered themselves. Consequently a minute amount of a trace element, by being used over and over, can regulate the breakdown or synthesis of biological material, or the transfer of energy, on a substantial scale. Hormones are chemical messengers, usually produced in one part of the body and used to control processes occurring in another. A review of the properties of the elements known (or suspected) to be essential will illustrate some of these matters.

Iron

Iron acts both as a structural element and as a trace element. The adult human body contains about four grams of iron; most of this is in haemoglobin, which gives red blood cells their colour and is responsible for the transport of oxygen from the lungs to the rest of the body. A small amount of iron is contained in myoglobin, used to store oxygen in the muscles, and minute amounts are involved in the action of various enzymes.

Iron is essential for the life of all organisms, except some bacteria. Empirical uses have been attributed to the healers of ancient times. Pale people were advised by the Chinese physicians of the second millenium BC to take pig's liver and blood; the Hippocratic prescription is said to have been rusty water.

Iodine

The maintenance of life in any organism depends on a multitude of chemical processes. Most elements perform a variety of functions; iodine is unique in having only one role, expressed by its presence in the thyroid hormones, which are essential for normal physical and mental development, and are responsible also for regulating the body's metabolic rate—that is, the rate at which energy is used. Iodine is unique also in the extent to which it is concentrated in one tissue. The total iodine content of the human body is 15–20 mg, of which 70–80 percent is found in the thyroid gland, a small organ in the neck.

The recommended daily intake of iodine in man is 150 micrograms (µg). The principal sources in normal diet are fish, leafy vegetables and dairy products; a small amount is usually present in water. Thyroid deficiency in man leads to goitre, a condition which is easily recognised. When the dietary intake is too low, the thyroid gland becomes larger, in the effort to make enough thyroid hormone. Goitre occurs, in man and animals, in regions where the natural iodine content of the soil or water is low. Substances present in cabbage, kale and other vegetables of the *Brassica* family can produce goitre in animals (even though the iodine content of the diet is adequate) by preventing the uptake of iodine by the thyroid gland or by interfering with the synthesis there of thyroid hormones.

Goitre caused by iodine deficiency is treated (or prevented) by the addition of iodine to the diet. This treatment (by seaweed or the ash from burnt sponges) was used empirically in many ancient civilisations long before the element itself was discovered in 1811. In many countries, table salt is now fortified with potassium iodide or iodate. In the United States and Canada, where table salt contains 75 PPM of iodine, this reinforcement alone is enough to provide the recommended daily intake. In Britain, the iodine content of salt is about 20 PPM.

There are several other sources of extra iodine. Factory-made bread often contains potassium iodate, which is added to improve the baking properties of dough; a single slice of bread may provide 150 µg of iodine. Erythrosine, a red dye used in colouring foodstuffs and pharmaceutical products, contains iodine to the extent of 58 percent. Iodine is found also in kelp tablets (made from seaweed) and other products sold in health food shops. Through these fortuitous additions, many people now have a daily intake several times greater than the recommended allowance. Harmful effects of excessive intake have been reported from some coastal regions in Japan, where the daily intake (largely from edible seaweed) exceeds 50 mg per day.

The toxicity of iodine depends on the chemical form in which it is taken. The use of potassium iodide in the treatment of syphilis was established early in the 19th century and continued to be recommended as late as the 1930s. The doses prescribed were very large—up to 20 grams per day for several months—but few ill effects were reported. Some patients developed skin rashes and others complained of running noses—a discomfort which, according to one eminent London physician, could be alleviated by doubling the dose. Potassium iodide or iodate tablets (in modest doses of 100 to 300 mg) are stockpiled today in the vicinity of nuclear power stations. Prompt administration reduces the uptake by the thyroid gland of radioactive iodine, a major constituent of fall-out from a nuclear accident.

DEATH ON SKID ROW

Doctors in Quebec City were baffled by the sudden increase in the incidence of heart failure. It had been known for nearly a century that alcoholics and other heavy drinkers were prone to heart disease—but the Quebec situation had the characteristics of an epidemic. 48 patients were admitted to hospital with established or incipient heart failure, in a relatively short space of time (between August 1965 and March 1966) after which no further cases appeared. Almost all of the victims lived, with no fixed address, in one area of the city. All were heavy drinkers of beer, most of them taking more than five litres a day of one particular brand. All of the victims were severely ill and 20 of them died.

The cause of the outbreak remained obscure until the provincial government—provoked by sensational newspaper reports claiming hundreds of deaths—appointed an expert committee to investigate the epidemic. Laboratory tests quickly showed that a virus could not be blamed. Attention then moved to the possibility of poisoning. Arsenic was suspected, for historical reasons (p 111) but analysis of hair clippings did not support this possibility. Further tests on the patients and the beer eliminated a long list of other toxic agents—some highly improbable—including radium, bismuth, cadmium, platinum, silver, insecticides and narcotics.

Most of the victims had been addicted to beer for many years and had not changed their drinking habits or their diets. The investigators knew that something must have happened to precipitate the epidemic. The clue came from the autopsy room. The thyroid glands of victims who had died showed changes similar to those seen in the rare cases of cobalt poisoning. It was found that cobalt had been added to the beer made in Quebec and consumed by the victims; but cobalt had been added to the beer of the same brand made in Montreal, where no cases of heart failure had been found among heavy drinkers.

Further enquiries uncovered a remarkable chain of events, beginning with improvements in hygiene in public houses. Newer detergents produced cleaner glasses, but left a thin film which was not always removed by rinsing. This film of detergent caused the collapse of the froth on the next pint of beer, to the annoyance of the consumer. Brewers in several countries welcomed the discovery (made and patented in Copenhagen in 1957) that the froth could be stabilised if cobalt (as the sulphate or chloride) was added to the beer. For reasons not explained, cobalt was added at a concentration of 0.075 PPM in Montreal and 1.2 PPM in Quebec.

It became clear that cobalt was the toxic agent sought by the investigating committee. The epidemic began within a month of the addition of cobalt to the suspect beer and ended within a month after the practice ceased—but doubts remained. The victims were only a small minority of those who drank large amounts of the suspect beer. The most enthusiastic drinkers had consumed only a few milligrams of cobalt per day—an amount well below the toxic threshold previously known; cobalt had been given in doses of more than 100 mg per day for the treatment of various diseases, where it did little good but no obvious harm. There was, however, no doubt that cobalt was the common factor associated with the damage in Quebec. The investigators could only suggest that the toxic effect of the element is enhanced by excessive consumption of beer and possibly by a diet low in protein. This combination is often found in heavy drinkers.

Similar epidemics were reported at about the same time from Omaha, Nebraska, where 11 out of 28 victims died, and from Louvain in Belgium, where only one of 24 patients died.

A USEFUL IMPURITY

Farmers in many countries were baffled for centuries by bush sickness. Horses grazed happily, but sheep and cattle wasted and died, unless they were moved to pastures which were healthier, though apparently no different. Investigations begun in New Zealand at the end of the 19th century produced convincing evidence incriminating iron deficiency. Affected animals were anaemic, the iron content of grass and soil was lower than in healthy areas, and salts of iron cured or prevented the disease. Similar findings were reported in other countries. For a while it seemed that the problem was solved. But more rigorous investigations (into bush sickness and enzootic marasmus, a similar condition affecting cattle and sheep in Western Australia) brought doubt and confusion on several counts:

(1) in parts of New Zealand the affected areas and the healthy pastures were little different in iron content.
(2) the therapeutic effect of iron compounds was not correlated with the amount of iron in the dose.
(3) the livers of affected animals contained ample stores of iron.
(4) whole liver was an effective remedy, even in doses which provided very little iron.

(5) an extract made from one of the popular remedies (limonite, an impure iron ore, also known as brown haematite) contained no iron at all, but was still effective.

It was suspected that an unidentified impurity was providing the benefit, rather than the iron salts in which it occurred. Limonite was found to contain appreciable amounts of nickel, but that was a false trail. Eventually cobalt was identified as the responsible impurity. Soil and grass in areas affected by bush sickness (as well as coast disease and some other afflictions of sheep and cattle) were deficient in cobalt, as were the livers of diseased animals; small doses of cobalt restored normal health.

COBALT IN HUMAN NUTRITION

Most of this work was done during the 1930s, but the nutritional role of cobalt was not fully understood at that time. Enlightenment came through the study (in human patients) of pernicious anaemia. This is a wasting disease characterised by failure of the normal development of red blood cells. Experiments made in the early 1920s by an American pathologist showed that raw or lightly cooked liver was effective in restoring normal haemoglobin levels in dogs which had been made anaemic by bleeding. In 1926, two Boston physicians found that large doses of liver were effective in treating pernicious anaemia—which had, until then, been incurable. The active component of the liver was eventually extracted in purified form in 1948 as vitamin B_{12}. Its red colour suggested the presence of cobalt, which was confirmed by further investigation.

Vitamin B_{12} is made by bacteria in the digestive tract, using cobalt present in the diet. In man, the process occurs only in the colon; absorption from there into the blood is too small to meet the body's necessary uptake—which is less than a twentieth of a microgram per day. In animals (such as sheep and cattle) which chew the cud, the necessary bacterial activity takes place in the rumen, where indigestible cellulose is made into a superior kind of silage. Only a small proportion of the vitamin produced in this way is absorbed, to meet the nutritional needs of the animals, but enough finds its way into food chains to satisfy human needs.

An adult needs a daily intake of about 5 µg of vitamin B_{12}, which is amply provided by a diet including normal amounts of meat, fish, eggs and dairy produce. It is not absorbed from the human digestive tract unless accompanied by the intrinsic factor—a substance of unknown composition made in the stomach. Patients suffering from pernicious anaemia are unable to absorb vitamin B_{12} from the diet, because their gastric juice lacks intrinsic factor. The vitamin is therefore given regularly by injection into a muscle. Except as a constituent of vitamin B_{12}, cobalt is not known to have any role in human nutrition.

Cattle and, more especially, sheep, use dietary cobalt very inefficiently in making vitamin B_{12} and are vulnerable to the diseases already mentioned if the soil and grass do not contain enough of the element. The deficiency can usually be corrected by slow release, over a period of years, from a suitably-formulated pellet placed in the rumen, or by adding a little cobalt sulphate to fertiliser spread on the soil each year. Horses, which make vitamin B_{12} by a more efficient process from the minute amounts of cobalt present in grass and hay, can survive on pastures which would not support sheep or cattle.

COPPER METABOLISM

Copper is needed by several enzymes in the body. The daily requirement has not been accurately established, since copper deficiency, though sometimes found in severely malnourished children, is virtually unknown in the human adult. From analysis of typical diets, and experiments on healthy volunteers, a daily intake of 2-3 mg has been recommended, though many people appear to remain healthy with smaller intakes. Copper bracelets have been popular for a century or more because of their supposed virtue in relieving rheumatic pain. Apart from a tendency to give the underlying skin a green colour they have no physiological effect—but may, of course, have a psychological effect on susceptible wearers.

The only significant disorder of copper metabolism in man is Wilson's disease. This is a rare heritable affliction in which copper accumulates in the tissues, causing damage to the brain and liver. It can now be treated successfully with penicillamine. This is a chelating agent—a chemical scavenger which removes unwanted metals from the tissues and holds them in a form suitable for excretion.

SWAYBACK AND MULTIPLE SCLEROSIS

Swayback, a paralytic disease of newborn and young lambs, was once common in many parts of Britain (where, in bad years, more than 70 percent of all lambs born were affected) as well as in Australia, Sweden and South America. Swayback lambs walk with a staggering gait or may be so seriously paralysed that they are unable to rise from the ground. When serious research began in the 1920s, an infectious cause was at first suspected, but was discounted when all attempts to transmit the disease from one animal to another were unsuccessful. In 1937 it was found that the disease could be prevented by adding copper salts to the diet of pregnant ewes.

Post-mortem examination of swayback lambs shows loss of myelin from parts of the brain. Myelin is a fatty substance which acts as an insulator surrounding bundles of nerve fibres in the brain and elsewhere in the nervous system. Its presence allows faster transmission of nervous signals and its absence results in disorganisation of communication within the brain or between the brain and other parts of the body. Loss of myelin is a characteristic of multiple sclerosis, a distressing human disease which has no known cause, though several have been suggested—including infection, climate, latitude and dietary intake of potatoes, animal fats, gooseberries or Brussels sprouts.

The possibility of a connection between the two diseases was sharply illuminated in 1947, when it was reported that four of the seven principal research workers in a British team studying swayback had developed multiple sclerosis; all four eventually died of the disease. This strange coincidence has not been explained—and nothing comparable has occurred elsewhere in the world. Treatment of multiple sclerosis with copper sulphate has been tried without success.

THE CASE OF THE PERSIAN DWARFS

'Mr Sherlock Holmes was sitting by the fire in his study at 221B Baker Street. His pipe was in his hand and his memorandum-book was open on his knee. "My dear Watson!" he greeted me, "It was good of you to come so quickly; your medical knowledge—and your military experience—may be serviceable in studying the mystery that we have on our hands." I protested that I was somewhat out of practice in both vocations, but Holmes continued: "The head of the Persian Department of the Foreign Office asked me to meet him at his club this morning. He was accompanied by our Ambassador in Teheran. They told me a most singular tale. It appears that a Persian physician recently made it known, very discreetly, in the souk of Shiraz that he wished to recruit a hundred dwarfs, for a secret mission lasting a year or more. They would be handsomely paid, housed in luxurious accommodation and would become famous throughout the world. "A strange tale indeed" I murmured, as Holmes went on: "He has now sought the assistance of the Army and has been spending much time in the company of a very senior General," "What devilish work is afoot?" I asked, "That, my dear Watson, is what we have to uncover." He stared into the fire for several minutes, tapped his pipe on the grate, and went on; "You recollect the case of the Bruce–Partington submarine?". I well remembered the incident in 1917, when the loss of secret plans threatened the stability of the British Empire. Holmes continued gravely: "The Admiralty fear that we may be facing another naval crisis—this time on the losing side. A defector from the OGPU has told Naval Intelligence that miniature torpedoes are being made in St. Petersburg," "You mean—" I began: "Exactly", Holmes interrupted. "Some fiendish treachery is afoot . . . a fleet of minature submarines may be let loose in the Persian Gulf . . . the Levant may go up in flames . . . another World War may be about to start . . . If we hurry, we shall catch the Orient Express before it leaves tonight . . .'

The Case of the Persian Dwarfs might have taxed the ingenuity of Sherlock Holmes—but the true story is stranger than anything from his casebook. It began in 1968, in the Nemazee Hospital, where Dr Ananda Prasad, a Visiting Professor in the Shiraz University Medical School, was looking after an unusual patient. This was a man of 21, with the appearance, height, weight and sexual development of a boy of 10. He was severely anaemic and had been unable to work, because of exhaustion and breathlessness, for four years. His diet consisted of unleavened bread with small amounts of milk and potatoes; he seldom had meat, eggs or vegetables, but had for many years been in the habit of eating clay. He was treated with an iron preparation and was discharged from hospital, considerably fitter though still underdeveloped, after ten weeks. Eight months later, now taking a more adequate diet, he showed further signs of physical and sexual maturity.

During the succeeding two years, Dr Prasad and his colleagues treated ten more patients with similar histories. It appeared—from their rapid response to medication and proper diet—that they had all been suffering from malnutrition and from iron deficiency in particular. But Dr Prasad suggested, with insight derived from his knowledge of animal nutrition, that the true cause of the patients' problems was zinc deficiency—a state not previously demonstrated in man, though known in animals since 1934. The importance of zinc in enzymes in the human body had been known since 1940.

In 1960 Dr Prasad moved to Cairo where, with better facilities for clinical investigation, he and his colleagues studied many dwarfs and showed that addition of zinc to the diet often produced rapid improvement in physical and sexual development. Other studies, in Egypt and in Iran, did not reproduce these findings, and many experts were reluctant to accept that zinc deficiency could occur in man. It was decided, by Dr Prasad's American and Iranian colleagues, that a definitive investigation should be made, in the hope of putting the issue beyond doubt. The organisation of the project, led by Dr James Halstead, was beset by many obstacles—not least the difficulty of finding enough young male dwarfs willing to spend a year as guinea pigs. It was expected that the recruitment process would be slow and tiring, since most of the possible subjects were likely to be found in the scattered villages where about 3 percent of 19–20 year old men were dwarfs; many would have to be excluded because of chronic afflictions such as tuberculosis or heart disease. The recruitment process was aided by the fortunate circumstances that the father-in-law of Dr Hossain Ronaghy (one of the collaborating physicians) was a high-ranking General in the Iranian army. Dr Ronaghy installed himself in one of the induction centres where conscripts were medically examined and interviewed more than 100 dwarfs who had been rejected by the army doctors.

The prospect of good food for a year, with entertainment and relaxation in a well-appointed house near the hospital, produced 15 suitable recruits, who were divided into three groups. All received a well-balanced diet for a year—except four who dropped out before the end of the study. One group received in addition a placebo pill each day. The second group received a daily pill containing 27 mg of zinc (as sulphate). The third group received a placebo pill during the first six months and a zinc pill during the remainder of the year. The results were convincing, despite the small number of subjects. The first group made modest progress in physical and sexual development, the second group made considerably greater progress and the experience of the third group was between these limits.

When these results were published in 1972, the status of zinc as an essential trace element in man was established, but a number of tantalising problems remained. The traditional diet of rural populations in Iran, Egypt and other Middle Eastern countries, with unleavened bread as a major component, is actually quite rich in zinc—more so than the typical diet of many healthy Europeans. The study of this paradox illuminates the important distinction between intake and uptake. The intake of a nutrient substance is the amount present in the food that is eaten. The uptake is the proportion (sometimes a very small proportion) of the intake that is transferred into the circulating blood—in other words, the biologically available fraction of the intake.

The uptake depends on several processes. In the stomach, food is exposed to hydrochloric acid and to pepsin, a digestive enzyme which hastens the breakdown of proteins. On leaving the stomach (from which very little is absorbed) and reaching the small intestine, the food enters an alkaline environment, provided by fluid from the pancreas, containing sodium bicarbonate and a number of enzymes, including trypsin. Most of the digestive processes take place here. Materials derived from the breakdown of food pass through the wall of the intestine and eventually reach the circulating blood.

It is now known that the uptake of zinc is greatly influenced by the presence of other substances in the diet. In particular, phytate—a constituent of many cereals—forms insoluble compounds in the stomach with dietary zinc, rendering it unavailable for absoprtion. Laboratory tests show that the zinc in foodstuffs with high phytate content should be completed precipitated. Yet human subjects absorb zinc normally from meals containing soya bean protein, which has a very high phytate content. The explanation is that other substances present in a mixed diet (containing meat, eggs and seafood as well as fruit and vegetables) can protect the zinc from precipitation by trapping it in soluble complexes. With a normal diet, the uptake of zinc is not greatly inhibited by the presence of phytate. Foods which are rich in fibre are usually rich in phytate, but wholemeal bread is a good source of zinc as well as of fibre.

SUFFICIENT YET DEFICIENT?

How much zinc do we need? The present recommendations of the US National Academy of Sciences prescribe a daily intake of 15 mg for an adult, 10 mg for a child under 11 years old and 20 mg for a pregnant woman. These figures are based on experimental evidence of a required uptake, in adults, of 6 mg per day and an assumption (for which there is limited evidence) that 40 percent of the intake is absorbed. It is generally believed that the actual daily intake in prosperous countries is close to the recommended level. This supposition was recently tested in Glasgow, with surprising results. Analysis of total diet samples showed a daily zinc intake of 10.1 mg for men and 7.6 mg for women. There are no obvious signs of damage from zinc deficiency in the population of the city, but further study showed that it is virtually impossible, in Glasgow or anywhere else, to reach the recommended intake. This situation arises because the official recommendation on zinc intake is incompatible with the recommendation (from the same authoritative source) on energy intake.

The daily energy intake recommended for adult males is 11.3 megajoules—equivalent to 2700 Calories in older and more familiar units. So a daily diet with 15 mg of zinc should be composed of food with an average of 1.33 mg of zinc per megajoule of energy. The measured ratio in the Glasgow diet study was 0.8. Table 3.3 shows that there is no realistic way of achieving the desirable increase to 1.3. In the foods which provide the bulk of any plausible diet, the ratio is much less than this. White flour contains only about 20 percent of the zinc present in wheat; so only compulsory use of unrefined flour in bread and other products would restore the balance between zinc and energy in the diet. The percentage loss of zinc in flour refining is about the same as for iron. But white flour is fortified with iron to make good the deficiency; perhaps fortification with zinc should be considered also.

Table 3.3 Zinc and energy content of typical diet (family of four) in the West of Scotland.

	Zinc mg/week	Energy MJ/week	µg zinc/kJ
meat	83	46.6	1.78
fried fish	2.4	3.6	0.67
cereals	58.2	99.9	0.58
dairy produce	47	67.6	0.69
vegetables	19.1	27.2	0.69
fruit, including tinned	2.3	8.5	0.26
soup	10.6	12.3	0.86
desserts	7.7	8.5	0.91
sugar confectionery	2.0	22.3	0.09
miscellaneous	3.5	3.7	0.96

(From Lyon, T D B, Smith H and Smith L B 1979 Zinc Deficiency in the West of Scotland? A Dietary Intake Study *Br J Nutr* **42** 413–416.)

Zinc is important in relation to drink as well as to food. The breakdown of alcohol in the liver requires an enzyme (alcohol dehydrogenase) in which zinc is an essential component. It has long been known that excessive consumption of alcohol causes damage to the liver. When this happens, the ability to detoxify alcohol is further impaired and more damage results. As long ago as 1957, Dr Bert Vallee and his colleagues in Boston found that in patients suffering from alcoholic cirrhosis of the liver the concentration of zinc was below normal in the blood and the liver, but above normal in the urine. In those with only moderate liver damage, function was improved by prolonged administration of zinc. It was also noted that the alcohol dehydrogenase activity of the liver was well above the normal in some subjects and well below in others. This finding is relevant to the common observation that some heavy drinkers escape liver damage.

Self-medication with tablets of zinc sulphate is often recommended by unofficial advisers on health care—both journalists and scientists. Zinc has been advocated as a treatment for anorexia nervosa, pre-menstrual tension, post-natal depression, tooth decay, AIDS, acne, schizophrenia and the common cold. Fortunately zinc is not particularly toxic in man.

AN IMAGINATIVE ACHIEVEMENT

Nerves are the pathways which conduct messages from one part of the body to another. Often (indeed usually) these messages are generated without conscious thought, by mechanisms which provide challenging problems for physiologists who like to know what is going on. In the early years of this century two of them—William Bayliss and Ernest Starling, of University College, London, were admiring a clever control system in the digestive tract. The breakdown of food in the small intestine, to provide the materials needed to maintain vital functions, needs enzymes made in the pancreas, an organ located behind the stomach. The pancreatic juice is used in an economical way, arriving where it is needed just as partly-digested food emerges from the stomach. Bayliss and Starling, working with animals, wanted to know how the signal reached the pancreas. Obviously, they thought, the arrival of food must act on nerve endings in the intestine to generate a signal for transmission to the pancreas. Which of the nerves associated with the pancreas was responsible? A simple way to find out was to cut the nerves one by one.

They cut them all—but the flow of pancreatic juice was still turned on and off as required by the delivery of food from the stomach. So how was the message being transmitted? In a remarkable leap of imagination, Bayliss and Starling looked for a chemical messenger. By 1902 they had found a substance, produced in the wall of the intestine, which stimulated the release of pancreatic juice in animals, even when they were not eating. The experimenters called it a 'hormone', after a Greek word meaning 'to activate'.

INSULIN AND DIABETES

Hormones, like enzymes, are proteins and, like enzymes, they depend on trace elements, either in their molecular structure or in close association. The discovery of the first hormone encouraged the search for others and eventually led to the discovery, in 1921, of insulin—an achievement to which millions of people owe their lives. The story of insulin goes back to 1889, when it was discovered in Germany that a condition very similar to diabetes could be produced in a dog by removing the pancreas. Diabetes is a disease (at that time invariably fatal) which is easily diagnosed by the presence of glucose in the urine.

Glucose, produced by the breakdown (in the digestive system) of dietary carbohydrate, is a major source of energy for the activities of cells and tissues. Its delivery to the cells where it is needed is regulated by a sophisticated control system. A small amount of glucose (about 4 grams—enough to supply they body's needs for only a couple of minutes) normally circulates in the blood. As it enters cells (the only places where it can do any good) the supply is topped up from the liver. When the amount of glucose in the blood increases after a meal, the pancreas releases insulin. This facilitates the removal of surplus glucose from the blood, for return to the liver, where it is changed into glycogen, a substance readily converted back to glucose to supply the energy needs of other tissues and organs.

In a diabetic patient, the pancreas does not produce enough insulin. Consequently glucose accumulates in the blood. It does not stay there; if it did, the blood would become more viscous, demanding extra work from the heart. The excess glucose in the blood of a diabetic patient is excreted in the urine and therefore serves no useful purpose. The continual loss of glucose, and the resulting disruption of the biochemical systems which keep the body alive, is potentially disastrous. Diabetes is now treated by insulin made from beef or pork pancreas. Advances in biotechnology have made possible the manufacture of synthetic insulin identical in structure to the human hormone—but no more effective than the animal product.

It is now known that diabetes cannot be controlled by insulin alone. It was reported in 1957 that, in rats fed on a highly-purified diet, containing all of the nutrients then known to be essential, blood glucose was not adequately controlled. This deficiency was successfully treated by giving brewer's yeast. The beneficial substance, designated GTF (glucose tolerance factor) was eventually found to be a compound of chromium, though its exact composition is not yet known. Inability to control blood glucose levels, even though insulin production is normal, has been found in a large proportion of malnourished children in Nigeria, Jordan and Turkey (though not in Egypt) and in adequately nourished middle-aged or elderly adults in the United States. Treatment by addition of chromium to the diet has sometimes been successful.

The daily intake of chromium believed to be safe and adequate is 50-200 µg, but even this small amount is not always achieved in countries (such as Finland) where the soil is deficient in chromium. Apart from brewer's yeast, thyme, cloves and black pepper contain relatively high concentrations.

SUCCESS SMELLS OF GARLIC

Dr Klaus Schwarz was baffled. His work on degenerative diseases of the liver, first in Heidelberg, then as a medical officer in the Wehrmacht, and now at the National Institutes of Health in Washington, seemed to have reached an impasse. Liver necrosis, produced in animals on diets low in protein, could be arrested by Factor 3—a substance of unknown composition, present in about a milligram of material obtained by laborious processing of a ton of pork kidneys. The active compound was so elusive that all attempts to isolate it in pure form were frustrated. In May, 1957 he recognised the vital clue—a smell of garlic from one of his preparations. This immediately suggested selenium, which was soon shown to be the essential element in factor 3. The study of selenium deficiency became an active field of research, intensified after it was established in 1973 that glutathione peroxidase, an enzyme present in liver, heart and other tissues in man and animals, contains selenium to the extent of four atoms per molecule. With this finding, the status of selenium as an essential element was put beyond doubt. Other aspects of the role of selenium in human health were vigorously explored; before long a major discovery was made in China.

SELENIUM AND HEART DISEASE

Heart failure is a common cause of death in middle-aged and elderly people—indeed, it is the last episode in everyone's life—but its appearance among children is unusual. One type of heart failure—sometimes acute, with mortality up to 50 percent in children, and sometimes chronic—had been known in China for many years before a severe outbreak in 1935 in Keshan County, from which the disease took its present name. Keshan disease occurred spasmodically among the peasant population in a wide area of mainly mountainous territory stretching from north-east to south-west China. Its distribution was irregular, with pockets of high incidence surrounded by regions relatively free from the disease. Local opinion suggested that a cause would be found in the soil. Selenium deficiency was suspected, and confirmed by measurements on soil, food, hair, blood and urine. The daily intake of selenium was found to be 10-20 μg; typical daily intakes in Europe and North America are 100-200 μg.

Beginning in 1974, a large-scale experiment to study the role of selenium in Keshan disease was made in Mianning County in south China. During a two year period, tablets containing sodium selenite were given regularly to more than 11 000 children under nine years old. A control group of more than 9000 children received placebo tablets, containing a trace of garlic to simulate the taste of the selenium salt. The results were convincing. Among the children who took the placebo tablets, 106 developed Keshan disease and 53 died. Among those treated with selenium there were 17 cases and one death. It was then decided to give selenium supplements to all children in the county, since a control group was clearly unnecessary; by 1977 the disease had been virtually eliminated.

Since 1976, selenium tablets have been issued more widely in the Keshan disease belt, with similar results. Sodium selenite is now added to table salt and sprayed on grain crops in the affected areas. The cause of the disease is not yet fully understood. The considerable fluctuation in incidence from year to year, and the seasonal peaks (winter in the north and summer in the south) suggest that selenium deficiency is not the only cause, though it is obviously the major influence on susceptibility to the disease.

Selenium is extracted commercially in the refining of copper ores, where it occurs as an impurity. Total production is about 15 000 tons per year. This relatively small output supports a wide variety of uses. Selenium sulphide is used in anti-dandruff shampoos and in ointments for the treatment of skin diseases in dogs. Other compounds are used in electrical rectifiers and solar cells. Small amounts of selenium are used in glass making to eliminate the green tinge produced by traces of iron.

Larger amounts are used in making red glass for signal lamps and traffic lights, and in pigments for paints and plastics. The dark glass increasingly used in windows or walls of buildings, to reduce glare and transmission of solar heat, also contains selenium compounds.

SELENIUM EXCESS AND DEFICIENCY

Selenium was not discovered as an element until 1817, but its effects were known long before then. Marco Polo commented on herbage, growing in Turkestan and Mongolia, which caused severe disorientation in grazing animals. This condition became well known in later centuries under the expressive name of 'blind staggers'. The most poisonous of the plants responsible was milk-vetch *(Astragalus)*, also known to the cowboys of the 19th century as locoweed. These plants have the remarkable property of concentrating selenium from the soil; one specimen found in Wyoming contained 1.4 percent of selenium. The accumulator plants, as they are called, suffer no ill effects, but are poisonous to grazing animals.

By 1960 it was known that grazing animals in many parts of the world were suffering from the effects of selenium deficiency. By then the element had a bad reputation. Scientists who advocated the addition of selenium to animal feed faced strong opposition, which was gradually overcome as governments allowed selenium supplementation on an increasing scale. Various methods are used—salt licks for free-ranging animals, fortified fertilisers and slow-release pellets (of selenium element compressed with iron filings, or incorporated in soluble glass) placed in the rumen, where they last for six months or longer.

Although there is no evidence of widespread selenium deficiency in human populations in the western world, there is a body of opinion that a substantial increase in dietary intake, to 300 µg per day, would protect against cancer. This belief draws some support from animal experiments showing that the incidence of tumours (either spontaneous or induced by exposure to carcinogenic chemicals) was reduced by adding selenium (as selenite) to diet or drinking water. Another study has suggested that the death rate from breast cancer in the United States is inversely correlated with selenium levels in grain and forage. Later studies in other countries have given conflicting results.

Sodium selenite tablets and yeast extracts containing other compounds of selenium are widely available in health food stores. In an experiment conducted in California in 1978, using healthy adult volunteers, it was found that organic compounds of selenium, incorporated in a specially-grown strain of yeast (indistinguishable from brewer's yeast) were absorbed to a small extent, as shown by increase in the concentration of selenium in the blood over a period of five weeks. Other tablets, sold as selenium bound to yeast or as selenium ascorbate, contained no absorbable selenium and were therefore without effect. Self-medication sometimes takes unexpected forms; people avid for selenium have been known to swallow sheep drench or other chemicals normally given to sick animals. Dietary supplementation (in absorbable form) is undoubtedly effective (in farm animals and in man) in the treatment or avoidance of severe deficiency. When the diet is already adequate, the beneficial effects of additional selenium (whether from the health food store or the veterinary pharmacy) are still debatable.

A COSTLY ERROR

Coincidences are, contrary to popular belief, quite common—and usually without much significance. But in 1927 Dr John Jones, a young general practitioner in Clydach, South Wales, was puzzled by the illness of two patients who consulted him in the space of a year. Both were found to be suffering from cancer of the ethmoid sinus (a hollow space in one of the skull bones near the nose). This is a rare condition, with an annual occurrence of about five cases per million people. Both patients worked in a nickel refinery. Dr Jones's suspicion of a carcinogenic hazard was reinforced when further cases were found among workers in the refinery, as well as an unexpectedly large number of cases of lung cancer, which was not the common disease that it has since become.

An error of judgement which was then made (not by Dr Jones) led to a number of deaths which might have been avoided. As no excess of cancer had been observed elsewhere in the nickel industry, attention was concentrated on one process peculiar to South Wales, involving the conversion of crudely-refined nickel to nickel carbonyl, $Ni(CO)_4$, and its subsequent decomposition to produce the metal in highly-purified form. Nickel carbonyl (a gas) had long been known to be poisonous, through its interference with respiration when inhaled; protective measures had been introduced in the industry before the end of the 19th century. The supposed association with cancer was a false assumption, which diverted attention from more hazardous processes (associated with inhalation of nickel dust) in refineries at Clydach and elsewhere. Occupational cancers of the ethmoid sinus and of the lung were virtually eliminated from the nickel refining industry in South Wales during the 1930s, but the hazard remained in other countries for many years later.

Nickel is now known to be an essential element in plants and in several animal species. As it occurs in a number of enzymes, it is presumed to be essential also in man. Impaired growth is observed in goats, rats and miniature pigs fed on diets containing only a few PPB of nickel; the effects may not be apparent until the second generation. since the deficiency in the maternal diet can be made good by depletion of the body stores. Nickel deficiency is unknown in man. Typical diets provide 300-500 ug per day; the minimum requirement is believed (by extrapolation from animal experiments) to be only about 20 ug per day. About 10 percent of women and 2 percent of men are allergic to nickel. Dermatitis from contact with nickel is often associated with the wearing of earrings, watchstraps and metal buckles—and is occasionally troublesome in people handling large numbers of coins. The remarkably high concentration of nickel in sweat—20 times the level in serum—is partly responsible for these discomforts.

MINOR ELEMENTS

Some trace elements have established roles and well-defined properties. Others have a collective significance rather like that of the Minor Prophets; most people have heard of them but they evoke little enthusiasm. Some are genuine enough but others should perhaps be relegated to the Apocrypha.

Viscount Wellington was probably preoccupied with the conduct of the Peninsular War when a case of tinned beef was delivered to his headquarters in 1813. The British firm of Donkin and Hall, who had established the world's first cannery in London, sent samples of their products to senior officers of the army and navy for appraisal. Wellington's secretary replied with a somewhat dubious recommendation; his Lordship had found the preserved beef very good—but was unable to reply in person as he was indisposed. Tinned food has been popular ever since—so much so that, according to one authority, American food manufacturers often add tin to bottled asparagus in order to reproduce the flavour that housewives expect. In man, tin is absorbed from the diet to the extent of only 2 percent or less. Although the intake may be appreciably increased by food or drink from unlacquered cans, poisoning from this cause is almost unknown.

The status of tin as an essential element is doubtful. Schwartz reported improved growth (though not up to normal levels) in undersized rats suffering from deficiency of vitamin B_2 (riboflavin). Other workers were unable to find any depression in growth rate when rats with an adequate supply of the vitamin were deprived of dietary tin.

Vanadium

Vanadium is widely distributed through the environment, though at very low concentration, from its occurrence in fuel oil (particularly of Venezuelan or Iranian origin) and in coal. It occurs in most foodstuffs at concentrations in the PPB range. Although it can be estimated with good sensitivity by modern analytical techniques, little information is available about its abundance in normal diets; the average daily intake is certainly less than 1 mg. Impaired growth (in rats and chickens) and reduced fertility (in rats) have been demonstrated using diets deficient in vanadium. The value of vanadium in protecting dental caries has been the subject of conflicting reports.

Essential Poisons?

Arsenic is well known as a poison, but is less familiar—and, to many people, less believable—as an essential trace element. It does, nevertheless, satisfy some of the prescribed conditions for essential status. In various organic compounds given at relatively large doses (50 PPM in feed) it controls disease and enhances weight gain in poultry and pigs. Danger to consumers is avoided by discontinuing the dietary supplementation a week before slaughter, to allow the arsenic level in the tissues to fall to an acceptable level. When the arsenic content of the diet is less that 50 PPB, signs of deficiency (including increased mortality, reduced growth rate, reduced milk yield and impaired reproductive performance) are observed in rats, goats and miniature pigs, and, more especially, in the offspring of female animals.

Cadmium (p 73) and lead (Chapter 5) have well-documented toxic effects, but may also be essential elements. Impairment of growth has been observed in a small number of animals fed on diets deficient in one or other of these elements, but further work will be needed to determine whether they are essential.

An Abundant Trace

Silicon is the most abundant element in the Earth's crust and in soil.

Until recently its biological effects had been studied only in relation to lung diseases caused by inhalation of the oxide (silica) or of asbestos, which contains silicates. Studies during the 1970s, based on comparison of growth rates in chickens and rats fed on highly-purified diets with and without supplementation, established the essential role of silicon in growth rate and in bone formation in these animals.

Molybdenum

Molybdenum is found, at very low concentrations, in many body tissues, both in man and in other animals. It is a constituent of several enzymes. It is certainly essential, but the minimum dietary requirement is not known with any accuracy, as balance studies have given conflicting results. Molybdenum deficiency has been induced in experimental animals fed on highly-purified diets, but is unknown in man, or in animals living under natural conditions. Studies during the 1960s, on experimental animals (by dietary manipulation) and in human populations (by epidemiological techniques) suggested that molybdenum conferred a measure of protection against dental caries. Later work did not confirm these results.

THE FLUORIDE STORY

The discovery that the incidence of caries can be greatly reduced by the addition of fluorine (in the form of a fluoride) to public water supplies provoked a long and remarkable campaign of misinformation.

The public perception of trace elements is related more often to their toxic effects than to their beneficial properties. The histories of arsenic, lead and mercury exemplify the common belief that chemical substances can be classified unequivocally as poisonous or harmful. This belief finds expression in the familiar assertions of people who ought to know better (and sometimes do) that the only safe dose of *X* is zero. The fact that *X* has always been present in the environment in huge amounts which are regularly changed (by natural mechanisms) from one location or one chemical form to another—or that a substance which is poisonous in large doses may be useful (even essential) in smaller doses—makes little impact on the popular understanding. In recent times the most striking demonstration of these issues has been provided in the campaign mounted in many countries to frustrate a major achievement in preventive medicine—the fluoridation of public water supplies.

This story begins, like so many told in the preceding pages, with an accidental discovery and with one man who saw how to make something of it. In the early years of this century Dr Frederick Mackay, a dentist in Colorado, noticed that a few of his patients had small brown stains on their teeth. The same discoloration was found in Texas and the condition became known as 'Texas teeth'. Dr Mackay noticed that stained teeth, whether in Colorado or Texas, hardly ever decayed. Continuing his enquiries, he was able to show that both the staining of teeth and the protective action against decay was attributable to something in the water. In 1931 this something was found to be a fluoride—that is, a salt of fluorine. Fluoride is present in almost all water supplies, though often at very low levels; water in the affected areas of Colorado and Texas contained quite large amounts.

The obvious question then to be considered was whether smaller amounts of fluoride in drinking water would produce stronger teeth without the brown stains. By 1943 it was established that in towns where the drinking water contained fluoride to the extent of one PPM children's teeth showed very little decay and no staining. Children are more vulnerable than adults because the process of decay (caries) rots the hard outer layers and begins in early life while the teeth are being formed.

The next question was whether the beneficial effect could be produced by adding fluoride to the reservoirs in areas where the natural level was below one PPM. Experiments in many parts of the world gave the same result. Wherever the fluoride level was adjusted to about one PPM the amount of tooth decay was reduced. In a few places where fluoridation was discontinued, because of local opposition, the prevalence of decay went back to its previous level.

Almost as soon as the fluoridation of public water supplies began in earnest, during the late 1940s, vigorous opposition appeared. The objections were of three kinds. First, the benefits were disputed, usually by considering children who were not in the vulnerable age groups and therefore did not show much benefit. Secondly, it was claimed that cancer and many other diseases were caused or aggravated by adding fluoride to water supplies to bring the concentration up to one PPM. The objectors were not influenced by the fact that these disasters do not occur in places where the water has a natural fluoride level of one PPM—nor do they occur in people who drink tea, which commonly contains two PPM of fluoride. A supplementary argument was that fluoride in large amounts was poisonous; this is true but irrelevant.

Thirdly, it was claimed that fluoridation was mass medication, undermining the liberty of the subject, this is an argument of only theoretical interest; all but the most primitive water authorities add 20 or 30 chemicals to their supply to deal with germs and other unwanted inclusions.

A SUBTLE JUDGEMENT

The anti-fluoridation movement was well supported financially and made some progress during the 1960s and 1970s before reaching its climax and downfall. In 1978 the Strathclyde Regional Council were moved by the appalling state of children's teeth in the West of Scotland (where the water supply was virtually free of fluoride) to raise the level artificially to one PPM. By exploiting the resources of the legal aid system the anti-fluoridation movement was able to challenge this decision in a trial of inordinate length and to bring witnesses from many parts of the world, at the expense of the British taxpayer. After the most comprehensive debate ever held on the issue, the judge decided that the scientific evidence levelled against fluoridation was worthless. Although on a narrow point of law he granted an interdict restraining the Regional Council from treating the water supply as they had wished, the objectors could not appeal against the judge's withering dismissal of the scientific and medical evidence opposing fluoridation. The point of law on which the judgement rested was overcome by subsequent legislation which opened the way for fluoridation in Strathclyde and in other localities where the outcome had been awaited.

The beneficial effects of fluorine on dental health do not constitute proof that the element is essential. Evidence on this point was obtained by Schwarz, who showed in 1972, by experiments using highly-purified diets in an ultra-clean environment, that fluorine is necessary for normal growth in rats.

A few more minor elements should be mentioned. Boron is essential for higher plants, but there is no evidence of a biological role in animals. A report, published in 1949, that lack of barium impeded growth in rats and guinea pigs has not been confirmed. Attempts to achieve a diet so low in aluminium as to provoke symptoms of deficiency in rats have not succeeded. Lithium is not known to have any function in living organisms. It is used, as the carbonate, in the treatment of depressive illness. The dose required is thousands of times greater then the normal dietary intake of about 100 µg per day and the mode of action is obscure.

The limits of analytical sensitivity and of biological insight have not been reached; it must be expected that some of the minor trace elements will in time prove to be more important than they now appear.

TRACES BY TUBE

Highly-purified diets are useful in studying the need for trace elements in animal nutrition and have a specialised role in human nutrition, when the normal mechanism for the uptake of dietary constituents from the alimentary canal is unavailable or inadequate. This difficulty occurs in several situations, for example:

(a) after surgical removal of part of the bowel, when the surface through which absorption takes place may be inadequate.

(b) After serious burning injury, when the patient's energy requirement is greatly increased—sometimes by as much as 100 percent—and cannot be met by normal feeding.

(c) In babies awaiting surgical correction of a congenital malformation.

In these circumstances it is necessary to by-pass the alimentary canal and to supply concentrated nutrients into the circulation directly. This technique is known as parenteral nutrition. Access is usually obtained by a catheter placed in the superior vena cava—a large vein (reached through the chest wall) which delivers blood (collected from smaller veins in the upper part of the body) to the heart, for reoxygenation in the lungs and redistribution through the arteries.

The major constituents of the nutrient solution are amino acids, vitamins, glucose (or dextrose) and structural elements, such as sodium, potassium, chlorine, calcium, phosphorus and magnesium, in suitable chemical forms. It is important also to supply trace elements, but it is difficult to know how much should be added. For many trace elements, the body's basic need when normally nourished is not accurately known. In a patient whose metabolic processes have been seriously disturbed, the uncertainty is greater.

Typical estimates of trace element needs during parenteral nutrition are given in table 3.4. These amounts are less than the familiar recommended dietary allowances of the elements concerned, because they are delivered into the circulation directly and are therefore intended to approximate to necessary uptake—not intake.

Table 3.4 Trace element requirements during parenteral nutrition.

Element	Daily requirement, μg
zinc	3200
iron	1200
fluorine	900
manganese	400
copper	300
iodine	130
chromium	55
selenium	35
molybdenum	20

HOW MUCH IS ENOUGH?

How much of an essential element is essential? In other words, how can we estimate the minimum daily requirement for a human subject? There are three methods suitable for general use, but none gives a wholly reliable answer. The first is to extrapolate from results of animal experiments. The differences in dietary constituents and metabolism between man and domestic animals are not so great as to make this approach completely useless. The minimum daily requirement of a particular element to maintain normal health in an animal can be estimated by performing experiments which would not be possible with human subjects. The requirement per kilogram of dry diet, or per kilogram of body mass, is taken as an estimate of human requirement. This gives a very rough estimate, which raises more questions than it answers. The amount required by an animal depends on many factors, including its species and stage of maturity. Normal health can be defined in relation to many observable processes, including growth rate (for young animals), loss of weight, blood level of the element in question and reproductive performance; the estimated requirement depends on which of these criteria is applied.

The second method is the metabolic balance study. Human volunteers, kept in a hospital, are supplied with a balanced diet containing an accurately known amount of the element under investigation. The daily loss, in urine and faeces, is measured over a period of a few weeks. A negative balance, in which the loss exceeds the intake, suggests that body stores are being drawn on because the intake is inadequate. Conversely, a positive balance may be taken to indicate that the intake is more than sufficient. It should be appreciated that the surplus may not be absorbed, for transfer to storage sites, but may simply pass out with the faeces. In that event, intake and output will be in balance over a range of intakes above the minimum requirement.

Table 3.5 Trace elements of interest in the biological environment.

	Estimated human body content	Daily intake in man			
		Observed	Recomm.	Believed safe & adequate	Max acceptable
boron	48 mg	1–3 mg			
fluorine	2.6 g	2–3 mg		1.5–4 mg	
silicon	18 g	1.2 g			
vanadium	100 µg	<1 mg			
chromium	6 mg	100 µg		50–200 µg	
manganese	12 mg	2–9 mg	2.5–5 mg		
iron	4 g	10 mg			60 mg
cobalt	1 mg	300 µg			
nickel	1 mg	400 µg			
copper	70 mg	2–3 mg		2–3 mg	35 mg
zinc	2.3 g	14 mg	15 mg		70 mg
arsenic	10 mg	100 µg			140 µg
selenium	10 mg	50–200µg		50–200 µg	
molybdenum	10 mg	150–500µg		150–500µg	
cadmium	50 mg	50–150µg			70–80 µg
tin	6 mg	0.2–4 mg			1.4 g
iodine	20 mg	64–700µg		50–1000µg	
mercury	13 mg	10–60µg			50 µg
lead	120 mg	300 µg			500 µg

Balance studies give widely discordant results. One study showed that the daily intake of selenium needed to achieve balance in North American adults was 80 µg for men and 57 µg for women—that is, about 1 µg per kg body weight. Other investigations gave figures of 24 µg per day in New Zealand women and 9 µg per day in Chinese men. It may be presumed that adaptation to lower intake develops in populations where (as in New Zealand and parts of China) the selenium content of soil and diet is low.

The third method is based on observation of dietary intakes in areas where the population appears to be healthy, and in other areas where signs of deficiency are apparent. Here too the results obtained are subject to much uncertainty, because intake does not always give a reliable indication of uptake (p 55).

Estimates of requirements and typical intakes for several trace elements are given in table 3.5. The many blank spaces in this table indicate the limited extent of present knowledge.

INTERACTIONS AMONG TRACE ELEMENTS

It is convenient to consider the biological role of the trace elements one by one; but many of their most interesting—and puzzling—effects are associated with interactions among them. Animals fed on a diet deficient in copper show characteristic signs, including anaemia, disorders of bone growth, changes in hair, wool or fur, and heart disease. But the converse is not always true. An animal may show signs of copper deficiency when its intake is fully adequate by normal standards. The explanation of this apparent paradox is that uptake does not always correspond to intake. The uptake of a trace element (that is, the amount transferred from the digestive tract to the circulating blood) may be greatly affected by the presence or absence of other elements in the diet. The uptake of copper may be impaired by the presence of small amounts of molybdenum or cadmium. The uptake of copper in ruminants and rats is impaired by high intakes of zinc; in pigs, high intake of copper leads to inadequate uptake of zinc.

Interactions of similar kinds are numerous, and are still being discovered more quickly than the underlying mechanisms can be explained or understood. Most of what is known about trace element interactions comes from observations or experiments on animals with little or no choice of diet. Grazing animals are susceptible to trace element imbalances resulting from localised geochemical anomalies. In prosperous countries, human food, because of its variety and widespread geographical origin, is not so vulnerable.

Because of biochemical differences among species, caution is needed in applying the findings of animal experiments to human populations—but some lessons are clear enough. It is known that copper is necessary (in man and other mammals) for the effective utilisation of iron—hence the anaemia associated with copper deficiency. It is also known that lack of iron or calcium in the diet conduces to increased uptake of lead; it is for this reason that the harmful effects of lead in the environment are exacerbated in undernourished children.

OUCH-OUCH!

There are many other examples of the enhancement of trace element toxicity by secondary effects. The Japanese Association of Orthopaedic Surgeons were told at their meeting in 1955 of a disease characterised by excruciating pain in the spine and legs, and believed to be caused by excessive exposure to cadmium, which is a common pollutant in localities where metal ores are mined or smelted. This condition, usually known by its Japanese name, 'itai-itai' (in English, 'ouch-ouch'), is caused by loss of calcium from the skeleton and is therefore more severe if the dietary intake of calcium is inadequate. Further studies in Japan showed that the cadmium intake of the victims described in the first reports was probably not enough to cause the disease. A condition indistinguishable from 'itai-itai' has been associated with excessive exposure to methyl mercury—for example, at Minamata (p 84). It is not clear that mercury alone was responsible, but there are as yet no clues to the other factors which may be involved.

Not all trace element interactions are harmful. In animals—and perhaps in man—supplementation of the diet with modest amounts of selenium protects against the toxic effects of methyl mercury or of cadmium. In one experiment, rats given a diet containing 10 PPM of methyl mercury died within six weeks, but similar animals on a diet that also included 0.5 PPM of selenium survived. In mercury miners, elevated levels of selenium and mercury have been found in brain and other tissues many years after the cessation of exposure to mercury. It seems possible that selenium and mercury combine in a form which is unavailable for further use, so reducing the toxic effects of both.

Some inferences relevant to human health, drawn from animal experiments, are still speculative—but should nevertheless be considered. Copper deficiency in animals produces a condition with characteristics similar to those associated with ischemic heart disease in man, including increased blood cholesterol, damage to arteries and heart muscle, abnormalities of the electrocardiogram and sometimes sudden death. It has been suggested that a high ratio of zinc to copper in the diet increases the risk of ischemic heart disease—and that a significant proportion of the population in the United States and other industrialised countries may not achieve the necessary daily intake of 2 mg of copper.

The multiplicity of interactions among trace elements has led to the suggestion, not entirely facetious, that every element interacts with every other. Sometimes in the history of science a vastly complicated mass of experimental data has been reduced to an orderly state, facilitating spectacular further progress, by the arrival of a simplifying concept—such as the Bohr atom or the double helix. The prospect of a comparable achievement tantalises many who have become fascinated by trace elements research.

ESSENTIAL OR FORTUITOUS?

The process by which a trace element is proved to be essential is usually long and tedious, including false starts, misleading clues and difficult experiments leading to ambiguous conclusions. An alternative approach developed recently in Glasgow will be described only in outline, as it uses mathematical techniques beyond the scope of this book.

The starting point is the observation that an element known to be essential is always under effective homeostatic control in the tissues where its functions are exercised. Any surplus amount (derived, for example, from the diet) is promptly stored elsewhere in the body or excreted. If the concentration falls, the deficiency is made up by drawing on the store.

Mechanism of this kind is remarkably effective. In rats, the dietary intake of manganese can be increased by a factor of 200 before tissue levels are doubled. If the concentration of an essential trace element in a particular tissue is measured in a large number of samples from healthy people, the values found will be fairly close to one another, with as many above the mean as below it; in technical terms, they will be normally distributed.

The refined control systems which stabilise the internal environment are not wasted on elements which perform no useful function. Measurements of the concentration of such an element will show a very wide range with an appreciable number of values much greater than the mean—in technical terms a skew distribution. In the tissues of a person exposed to abnormal amounts of the element, the concentration goes on rising. Partial removal of hair, skin or nail may ease the situation but the concentration in blood or other vulnerable tissues will continue to increase, even to the level at which toxic or fatal effects develop.

Measurement of the concentration of an element in the tissues of a large number of healthy people forms the basis of a test to decide whether or not the element is essential. If the concentration values are normally distributed, the element is probably essential; if the distribution is skew, the element is probably not essential. It is possible to devise a numerical scale of skewness and therefore of the essentiality of a particular element.

We know (p 72) that there are many interactions among trace elements; the signs of excess or deficiency of one trace element can be influenced by the presence or absence of another element in the diet. So perhaps the concept of essential and non-essential trace elements is an over-simplification. Perhaps we should think in terms of a continuous spectrum of essentiality, in which the conspicuous extremes have already been identified. There is some evidence that an element may be more essential in some tissues than in others. Strontium behaves rather like an essential element in bone—where it can, to a limited extent, fulfil a similar role to calcium. In cartilage, a tissue which contains no calcium, strontium behaves as a non-essential element.

CHAPTER 4

Mercury and Venus

A ROYAL DENTIST

James IV, King of Scotland from 1488 until his death in 1513 at the Battle of Flodden, was a man of wide learning. He had a considerable interest in dental health—beginning, reasonably enough, with his experience as a patient. In the accounts of the Lord High Treasurer for 1503 we read of a payment of 14 shillings.

'to the barbour that cam to tak furth the Kingis tuth'.

By 1511 the King had learnt enough to go into practice for himself—with unusual financial arrangements. A further payment of 14 shillings was made in that year

'to ane fellow because the King pullit furth his twtht'.

A few days later the Lord High Treasurer paid out again

'to Kynnard a barbour for twa teith drawin furtht of his hed by the King'.

Before the 19th century, decayed teeth were usually repaired, if at all, by packing pellets of gold foil into the cavities. This technique can be quite effective, for gold is easily welded under pressure, even at low temperatures, but the process is slow and painful. Had the king been more enterprising he might have advanced the progress of dentistry by three centuries. The resources were certainly available in the laboratory occupied by the royal alchemist, John Damian.

This colourful character was, as we know from other references in the Lord High Treasurer's records, well supplied with the basic needs of his craft, including gold, silver, mercury and strong drink. Since the conversion of mercury into gold was one of the aims of the alchemists, Damian would certainly have known that mercury has the ability to dissolve gold, silver or other metals, forming a solution, known as an amalgam, which is liquid at first but soon hardens. Gold amalgam was used as early as the first century AD for the decoration of metal objects. The amalgam was applied as a liquid and then heated to drive off the mercury, leaving a thin film of gold—and releasing mercury vapour, which was recognised as a serious occupational hazard.

The royal dentist and his alchemist did not make the necessary leap of imagination. The first amalgam used in dentistry was made in 1818 from bismuth, tin, lead and mercury. A material composed of mercury and silver filings was introduced in the following year. These crude amalgams were not always effective; fillings expanded so much while setting that teeth sometimes broke. An amalgam of copper and mercury, first made during the 1850s, was no more successful, but better formulations were evolved later in the century. The amalgam commonly used today is made from equal parts of mercury and an alloy containing silver (not less than 65 percent), tin (not less than 25 percent), copper (not more than 6 percent) and tin (not more than 2 percent). Some dentists prefer to use a higher proportion of mercury; if this is done, the excess mercury has to be squeezed out before the filling material is inserted in the tooth cavity.

Modern amalgam is a remarkable material. Its thermal and mechanical properties are close to those of the tooth substance that it replaces; a filling withstands exposure to hot liquids or iced drinks without causing discomfort, and wears down at the same rate as the surrounding tooth. These properties could not have been predicted. A 19th century chemist, asked to produce a material for filling holes in teeth, would probably have tried to make a cement based on calcium phosphate, which is a major constituent of bone. Twentieth century chemists have devised various plastics and synthetic resins—but amalgam still holds the field.

When amalgam first became available in the dentist's surgery it was widely proclaimed (in the United States, though not in Europe) that the new material was dangerous to patients. The Amalgam War, which divided the American dental profession, continued for most of the 19th century.

The hazards of occupational exposure to mercury were well known. Early in the 18th century, Ramazzini wrote:

'We all know what terrible maladies are contracted from mercury by goldsmiths, especially those employed in gilding silver and copper objects. This work cannot be done without the use of amalgam, and when they later drive off the mercury by fire they cannot avoid receiving the poisonous fumes into their mouths, even though they turn away their faces. Hence craftsmen of this sort very soon become subject to vertigo, asthma, and paralysis. Very few of them reach old age, and even when they do not die young their health is so terribly undermined that they pray for death'.

The use of mercuric nitrate in hatmaking, to convert animal fur into felt, was also recognised as dangerous. Whether the Mad Hatter in Alice in Wonderland was deranged from this cause is a matter on which opinions differ. It is also believed by some that the expression 'mad as a hatter' is a corruption of 'mad as an adder'. The tremor which is a characteristic symptom of mercury poisoning was certainly found among hatmakers and was often known as the Danbury Shakes, from the town in Connecticut which was an important centre of the industry.

In 19th century dentistry, the real hazard was to the reputations and purses of practitioners devoted to older and more lucrative methods using gold. The controversy died down, but was activated again in the 1920s by Alfred Stock, a German analyst. He found mercury in the urine of patients whose teeth had amalgam fillings, and concluded that they were in danger of poisoning. The levels that he reported were no greater than those found in people without amalgam fillings and did not justify alarm, since mercury is a normal constituent of the diet (p 87). Analysis of urine is, in any event, of little use in the assessment or control of mercury hazards. Even when mercury poisoning is all too obvious from other signs, urinary mercury levels fluctuate in a very irregular way and are not at all correlated with toxic symptoms.

It is now established that, in general, amalgam fillings present no hazard to patients. Metallic mercury and amalgam are virtually insoluble in saliva and in gastric juice and are therefore not absorbed to an appreciable extent from the gut. A rise in the excretion of mercury in the urine can be detected for a few days after the insertion (or removal) of amalgam fillings; the increase is small in relation to the normal mercury excretion. In recent years it has been claimed that unexplained ill health in many people is attributable to mercury allergy—and that removal of amalgam fillings restores the patients to fitness. The number of people who are genuinely allergic to mercury is very small indeed. For others, the benefit of amalgam removal can only be psychological.

For the patient, who may have only a few fillings in a year, there is, in general, no appreciable hazard from mercury amalgam. But for the dentist, who may insert 20 or more fillings in a day, the danger is greater. The materials for each filling must be prepared separately, so that the amalgam may be inserted into the tooth cavity before it has hardened. At one time the amalgam was prepared by grinding the components together with a pestle and mortar. The pellet was removed by hand and perhaps squeezed to remove excess mercury. Concern over possible hazards in these procedures encouraged the introduction of electrically-driven amalgamators, in which the necessary materials, contained in a plastic capsule, were combined by vigorous agitation. This process is not necessarily free from hazard, for the capsule becomes quite hot when it is shaken, and emits a puff of mercury vapour when it is opened.

The amalgam pellet must be pressed into the cavity, carved to reproduce the contours of the tooth material that it replaces, and polished. Since the dentist must inspect the tooth closely during these manipulations, it is virtually impossible for him to avoid inhaling some of the vapour which mercury emits quite abundantly at body temperature. When inhaled, mercury passes quickly from the lungs into the blood. Some is excreted, mainly in the urine, and some is stored in the kidneys. It is possible that some is converted to methyl mercury (p 83), a much more toxic substance which can make its way into the brain.

Despite the abundant evidence, gathered over many centuries, of the health hazards from inhalation of mercury vapour, the prevailing opinion, as late as the 1960s, was that dentists were not at risk from this cause. Earlier investigations had given negative or inconclusive results, largely because analytical techniques of sufficient sensitivity were not available. Activation analysis offered a way out of this difficulty and was used in a 20-year study which began in Glasgow in 1964.

This investigation, which eventually involved nearly 1000 dental workers in Scotland, was based on the measurement of mercury in four small samples provided by each participant—head hair, pubic hair, finger nail and toe nail. It was expected that contamination would be higher in head hair and finger nail (containing mercury delivered by two routes—internally, having been inhaled, and externally, from vapour and dust) than in pubic hair and toe nail, in which mercury would be deposited only by the internal route. When the analytical findings were studied, it was noticed with surprise that many participants (identified at that stage only by code numbers) had more mercury in toe nails than in finger nails. Most of these subjects were women, but the unexpected findings were not explained until visits to a few workplaces showed that many female surgery assistants wore open-toed sandals. Mercury vapour, which is much heavier than air, settled on the exposed toes as it fell to the floor (table 4.1).

Table 4.1 Mercury in toe nails, PPM: open-toed *v* covered footwear.

	Open-toed	Covered
number of samples	177	445
mercury, PPM, range	0.16-139	0.10-209
geometric mean	1.49	1.12

The survey showed that the average (geometric mean) levels of mercury in hair and nail samples from dental workers were significantly higher than in samples from members of the general public. A number of other findings emerged:

(a) Dental workers in general practice absorb considerably more mercury than those working in Health Service clinics (table 4.2). There are several possible reasons for this difference. In general practice, the dentist is paid for each item of service (extractions, fillings, etc). The clinic dentist receives a salary which is not directly related to his work load. The dentist in general practice provides his own premises and equipment, which are not always of as high a standard as in government-financed clinics.

Table 4.2 Comparison of mercury levels—general practice staff and health service clinic staff.

	Mercury, PPM, geometric mean	
	GP	Clinic
head hair	3.51	1.54
pubic hair	1.71	0.76
finger nail	12.1	2.65
toe nail	1.71	0.57

(b) Dentists who smoke tobacco absorb more mercury than non-smokers (table 4.3). Here again the explanation is not hard to find. The temperature of a lighted cigarette is about 900°C. At this temperature any mercury transferred from the fingers to the cigarette paper will be completely vapourised; the vapour may be inhaled and may be transformed (by chemical reactions inside the burning cigarette) into more toxic substances. This hazard is obviously greater in dentists who smoke during the working day, as was illustrated in the unhappy experiences of two dentists, each of whom was in the habit of smoking a cigarette in the interval between patients.

Table 4.3 Mercury in pubic hair: smokers *v* non-smokers.

	Mercury, PPM	
	Smokers	Non-smokers
number of subjects	141	473
range	0.33-174	0.09-67
geometric mean	1.62	1.24

'Mr A had been vaguely unwell for a year, suffering from headaches, muscle pain and sleep disturbance. He had lost 14 kilograms in weight. He felt better after a three week holiday, but relapsed two weeks after resuming work. Finding that alcohol relieved his symptoms, he indulged rather freely, generating suspicion that his health problems were caused by drink. When he began to exhibit finger tremor, irritability, loss of memory and visual disturbances, his colleagues suggested a diagnosis of mercury poisoning and he was admitted to hospital. Mercury levels in hair and nail were considerably above normal values.

Mr B complained, over a period of three years, of tremor, nausea, visual disturbance and loss of manual dexterity, he tended to stagger while standing or walking. His physicians were investigating the possibility of neurological or endocrine disorders; the patient himself suggested a diagnosis of mercury poisoning'.

There is no effective way of dealing with mercury poisoning. If penicillamine is given, excretion of mercury in the urine is increased. It is probable that this increase comes from mercury which is freely available, rather than that which is deposited in the brain or kidney, where the damage is done.

One fatal case of mercury poisoning in a dental worker has been recorded in recent times. In 1969 a female dental surgery assistant aged 42, who had followed her occupation for more than 20 years without serious ill health, developed a sudden illness with signs of kidney failure and died four weeks later. *Post-mortem* examination revealed no abnormalities, except in the kidneys, which contained mercury at a concentration of 520 PPM—about 100 times the normal level.

MERCURY AND ALCHEMY

The history of mercury goes back to the beginning of chemistry—indeed, to the beginning of alchemy. The earliest records describe the use of mercury in the second century BC by Chinese alchemists in their search for the elixir of life, which would confer immortality on its possessors. The Taoist religion, founded about 600 BC, was later influenced considerably by alchemical concepts. Taoists believed that all animate and inanimate objects were formed by the union of opposing principles—the Yin, which was cold, passive, female and based in the Earth, and the Yang, which was hot, active, male and derived from the Sun. Mercury was identified with the Yin and sulphur with the Yang.

These ideas developed further in the Muslim world under the inspiration of Geber (Jabir al-Hayyam), who was alchemist at the court of Harun al-Rashid, the 8th century Caliph of Baghdad. Geber taught that all metals were formed in the earth by the union of sulphur and mercury. If these two elements were of the highest purity—a state which the alchemists sought conscientiously but never achieved—the product was gold; otherwise, the union produced silver, copper, iron, tin or lead. The transmutation of base metal into gold was dismissed as impossible by some of Geber's contemporaries and by many of his successors, though the notion persisted long enough to be taken seriously by Isaac Newton (pp 9)

OCCURRENCE AND INDUSTRIAL USES

Mercury is one of the very few materials (caesium and gallium are the others) which exist as liquids at ordinary temperatures—hence its colloquial name, quicksilver and its latin name, hydrargum (liquid silver), from which its chemical symbol Hg is derived. Widely distributed in the earth's crust, usually as the sulphide (cinnabar), it is easily produced, at 99 percent purity, by heating the crushed ore and condensing the vapour. Annual production in this way is about 10 000 tons.

A major use of mercury is in the chloralkali process for the production of sodium and chlorine. A stream of mercury forms the cathode in the electrolytic decomposition of brine. Sodium appears at the cathode, forms an amalgam with the mercury and flows to the next stage in the process, where it is converted to caustic soda. Chlorine is recovered from the anode of the electrolytic cell and about 95 percent of the mercury is recycled. A considerable amount of mercury is used in making small batteries (for hearing aids, watches, cameras, calculators and other devices), mercury discharge lamps (for street lighting or the production of ultraviolet radiation), industrial power rectifiers, thermometers and other scientific equipment.

Many mercury compounds are effective in preventing the growth of fungi, algae and bacteria. For this reason they have been widely used in agriculture and in the making of leather, paint, wood pulp and paper. Since the 1960s these uses have been decreasing, because of concern about pollution of the environment and associated dangers to human and animal life (p 84). Mercuric chloride is used as a catalyst in the production (from acetylene) of vinyl chloride, the starting point for the manufacture of the ubiquitous PVC (polyvinyl chloride). Mercuric sulphate is used as a catalyst in the production, also from acetylene, of polyvinyl acetate, a constituent of many emulsion paints. About 400 tons of mercury are used each year in making dental amalgam (p 77).

The world's annual production of mercury doubled, to about 10 000 tons, in the early 1950s, to meet increased demand from the agricultural industry, mainly for pesticides to control the growth of fungi on seed grain. A number of fungi attack seeds, grow into the grain as it develops and make it worthless. Seed dressing by chemical treatment, before sowing, to reduce fungal attack is a long-established practice. In his *Essays on Field Husbandry in New England* (1748) Jared Eliot declared:

'If a farmer in England should sow his Wheat dry without steeping in some proper liquor, he would be accounted a bad husbandman'.

Sulphur, copper sulphate, arsenic trioxide and many other chemicals were used in the 19th century. Mercuric chloride (corrosive sublimate) was introduced in 1890, but organic compounds of mercury became popular in the present century. Fungicides based on methyl mercury were available from 1930 and were found to be very effective. Demand for these fungicides increased greatly in Europe during the years after

1945, when governments and farmers were concerned over food shortages.

Large quantities of seed wheat and other grains were treated, apparently cheaply and effectively, before distribution to farmers. Serious outbreaks of illness in Iraq (1956 and 1960), Pakistan (1969) and Guatemala (1963–65), characterised by loss of co-ordination, slurred speech, loss of hearing and vision and, in many cases, coma and death, were diagnosed as poisoning from organic mercurials. The crop from treated seed contains negligible amounts of mercury; but if the seed is made into flour, or fed to livestock, to produce food for human consumption, the consequences can be catastrophic, as they were in the 1972 disaster in Iraq.

73 000 tons of seed wheat and 22 000 tons of seed barley were imported late in 1971 and distributed during the following three months. Many farmers did not receive these supplies until after they had sown their own seed; some used as much as 100 kg of treated seeds in making bread. More than 6000 people were admitted to hospital suffering from mercury poisoning and nearly 500 died. Analysis of hair and blood samples gave much valuable information.

Methyl mercury is excreted rather slowly, the body burden falling by about 1 percent per day. By estimating the amount of contaminated bread eaten by patients before the onset of symptoms, and the amount of mercury in the blood on admission to hospital, it was possible to estimate the total body burden of mercury. It appeared that victims who had accumulated less than 200 mg of mercury did not die and that those whose bodies contained less than 25 mg suffered no harmful effects. A similar study of the poisonings in Japan suggested a threshold of 30 mg. These figures have since been used elsewhere in estimating maximum permissible levels of methyl mercury in food.

Methyl mercury was implicated in two environmental disasters which occurred in Japan during the 1950s. There the waters of Minamata Bay and the Agano River were polluted by organic mercurials and many other toxic chemicals released from factories. More than 50 people died in these outbreaks.

Increased awareness of the hazards of organic mercurials led to further investigations and precautions, especially in Sweden. Naturalists there reported dwindling populations of many seed-eating birds, such as pigeons, pheasants and partridges. Birds of prey, including some owls and eagles, were also affected. Suspicion fell on pesticides—not for the first time. *The Shooting Times* reported in 1884:

The Danger of Dressed Wheat

'An unusual conviction was recorded at Peterborough Police Court on Thursday against a farmer named Robert Goodfellow. The prosecution was conducted by the police, through the defendant having scattered dressed wheat on an already drilled and harrowed field, thereby causing danger to life, as numerous dead birds—woodpigeons, crows and a magpie—had been discovered on the defendant's and adjoining land. The prosecution took the view that danger to human life might result from the continuance of such a practice'

Mr Goodfellow was probably hoping to poison some of the birds which fed on his seeds.

When unusually high levels of mercury were found (by activation analysis) in feathers and other tissues from dead birds in Sweden, farmers were urged (without much success) to reduce the application of mercurial dressings to seeds. A puzzling feature was the observation that Swedish birds contained much greater amounts of mercury than those in Denmark, where organic mercurials were also used for seed dressing. Although activation analysis is a very sensitive technique for the estimation of mercury, it gives no information about the chemical form in which mercury is present. By the early 1960s, Swedish scientists had developed analytical techniques for the measurement of methyl mercury. It was then found that most of the mercury in Swedish birds was in this form. Danish farmers used other mercury compounds, which are rapidly broken down in birds and animals—in contrast to methyl mercury, which lingers in the body for a long time. In 1965 the use of methyl mercury in agriculture was forbidden in Sweden—an example which was gradually followed in other countries.

Meanwhile, monitoring of fish in Sweden and elsewhere was producing unexpected results of great significance. Scientists were puzzled because methyl mercury was found in fish from waters known to be free of pollution by organic mercurials. By the end of the 1960s it was established that almost all of the mercury present in fish—wherever they are caught—is methyl mercury. The explanation for this remarkable finding is that bacteria present in sediments at the bottom of rivers, lakes and the sea are capable of changing inorganic mercury into methyl mercury. As it rises through the water, some of this unwelcome contaminant is absorbed by fish; some reaches the atmosphere, where it is broken down by sunlight, eventually into methane and mercury.

THE MERCURY CYCLE

It is now clear that the movement of mercury in the environment is a cyclic process. Mercury vapour escapes continually from the earth's surface—both land and sea. Other inputs to the atmosphere come from volcanic activity and from the burning of fossil fuels. About half of the mercury produced for industrial use is recycled at or near the place of use; the rest is added to the environment. Mercury vapour is removed from the air in rain and eventually finds its way back into the sea, along with mercury carried down by rivers. Of the mercury that reaches the sea, most remains dissolved until it rises into the atmosphere again. The remainder goes into the seabed sediments, from which a roughly equal amount rises through the water—some in the form of methyl mercury. The amounts of mercury involved in these processes are very large (table 4.4).

Table 4.4 The mercury cycle: some annual transfers, tons.

into atmosphere: vapour from earth's surface — land	8000
— sea	22 000
human activity — roasting of sulphide ores	2000
— burning of fossil fuels	5000
— chloralkali process	3000
from atmosphere: rain	40 000
into sea: from rivers	5000
from rain	27 000
from sea: mercury vapour into atmosphere	22 000
mercury into sediments	3500

Though the figures given in table 4.4 are subject to large uncertainties, they illuminate some important points:

(1) the mercury content of the atmosphere is in a steady state.
(2) The input of mercury to the sea is greater than the loss to seabed sediments. The difference (at present 6500 tons per year) is insignificant in comparison to the total mercury content of the sea, estimated to be at least 40 million tons.

(3) The total amount of mercury processed by man since 1900 is no more than 400 000 tons. If all of this were added to the sea, the concentration of mercury there would be increased by only 1 percent. The pollution of rivers, lakes and coastal waters is, of course, a different matter.

MERCURY IN FOOD

In seeking to avoid hazard from potentially toxic substances in food, governments (and the national or international bodies which advise them) normally have two options—to prescribe a maximum allowable daily intake or to prescribe maximum allowable concentrations in particular foodstuffs. Occasionally, in moments of panic, the sale of particular foodstuffs may be forbidden. In practice, a maximum allowable amount or concentration is found by choosing a number somewhere between the level normally present and the lowest level known to be harmful. If there is enough room between these two levels, the maximum allowable daily intake may be set at a fraction (commonly one-tenth) of the threshold level for harmful effects. The difficulty of making judgements of this kind is well illustrated by the history of recent efforts to regulate dietary intake of mercury.

This issue came to the attention of the public in the late 1960s, when the alarm was raised by scientists and environmental pressure groups in Sweden and North America. Steps were taken to reduce mercury discharges from industrial activity. Fishing was banned in many polluted lakes and rivers and a limit of 0.5 PPM was prescribed for fish caught elsewhere. Concern increased in 1970, when an American scientist announced that a sample of tinned tuna fish contained 0.86 PPM of mercury. Further investigation found some samples with more than 1 PPM, and suggested that nearly 1000 million tins in shops and warehouses contained more than 0.5 PPM. Since the tuna (or tunny) is a deep-sea fish, it should have been obvious that it could not be contaminated by man-made pollution, which has a negligible effect on mercury levels in the sea. The media and the public were not noticeably reassured when it was shown, a year later, that the mercury level in recently caught (or canned) tuna was the same as in museum specimens between 60 and 90 years old.

In judging the possibility of harmful effects from toxic substances in food it is, of course, necessary to consider the total intake rather than the concentration in particular items of the diet. Estimates of total dietary intake are often uncertain, because of experimental difficulties. The present official estimate of mercury in the British diet—between 5 and 10 µg per person per day—is based on measurements made in government laboratories in 1970. The technique used (atomic absorption) was not the most suitable; in more than half of the samples analysed, no mercury was detected. A later survey in Glasgow, using activation analysis, examined more than 100 items of diet and found mercury in all but five—Coca-Cola, lemonade, butter, toffee and red Bordeaux wine. In this survey, the average daily intake of mercury was estimated at about 60 µg per person.

Food is not the only source of mercury in the body. Atmospheric contamination, mainly through the burning of fossil fuels, makes a significant contribution in many countries. A few people augment their intake by bizarre dietary preferences. When hair samples from a group of Canadian Indians were analysed in 1970, the highest mercury concentration was found in a 46-year old woman living in Manitoba, who suffered from headaches, dizzy spells and visual difficulties. Enquiry showed that she had a craving for paper. She regularly ate cardboard and cigarette packets and had once consumed a whole paperback book in a day. In these ways she absorbed significant amounts of mercury from the residue of the materials used to control the growth of mould on wood pulp.

Craving for substances not normally regarded as nutritious—such as paper, paint and clay—is a recognised medical condition, known as pica. It is believed to be a sign of iron deficiency. The Canadian patient, after a week's treatment with ferrous sulphate, expressed an aversion to paper and stopped eating it.

MERCURY IN MEDICINE

Mercury is one of the oldest remedies known to medicine. It was used (mixed with grease to make an ointment) in first-century Rome and was known (both as the metal and the naturally-occurring sulphide) to the physicians of ancient Egypt, Assyria and Babylonia long before that. Harmful effects were noted by Dioscorides, a Greek physician who served in Nero's army, during the first century, and by some of his contemporaries. Galen, a Greek physician who flourished during the second century, insisted that mercury was poisonous—a judgement which only slightly restrained the enthusiasm of prescribers in later times.

The first man-made compound of mercury to be used in medicine (often with disastrous results) was mercuric chloride, $HgCl_2$, a corrosive salt which was known as early as the tenth century. When metallic mercury, copper sulphate and common salt are heated together, the hydrochloric acid which is produced combines with the mercury. The product sublimes, passing directly from solid to vapour—hence the name corrosive sublimate. This preparation has been seen by some as marking the start of the systematic study of synthetic chemistry. Mercurous chloride, $HgCl$, usually known as calomel, is relatively insoluble and therefore less toxic than corrosive sublimate.

By the 16th century, when Paracelsus came on the scene with the first attempt to introduce scientific ideas into medical treatment, mercury was widely used, especially in the treatment of skin diseases, but was discredited among the more thoughtful physicians. Paracelsus taught that mercury was valuable if properly used; the correct treatment was not merely to administer mercury, but to administer it in the correct amount and chemical form. The novel doctrine that dosage had to be quantitative as well as qualitative was gradually accepted. Some of the other precepts taught by Paracelsus were, by the standards of later centuries, less than rational. He had much faith in the doctrine of signatures, which declared that Nature provided obvious hints for the observant physician. So, for example, liverwort and kidneywort are made with leaves in the shape of the parts that they cure. Syphilis, he explained, is a venereal disease, acquired from venal women; Mercury is the Roman god of commerce and the market place; therefore the element mercury is to be used in the treatment of syphilis.

MERCURY AND VENUS

Paracelsus claimed considerable success in the treatment of syphilis, which was in some respects the 16th century counterpart of AIDS. Mercury compounds were sometimes given by mouth, though the commonest route was through the skin. There were two techniques—inunction and fumigation, both of which survived into the 20th century. Inunction was achieved by rubbing the skin with mercurial ointment, sometimes containing corrosive sublimate as well as the metal. Fumigation was done in a powdering tub—a large wooden barrel, otherwise used for curing sides of meat. The patient, his skin well greased, sat naked in the tub, above a heated tray spread with sulphide or other mercurial compound. The vapour condensed on the grease and some of the element passed through the skin.

Rabelais, himself a physician, gives a vivid account of the early effects of inunction:

'. . . but what shall I say of those poor men that are plagued with the pox and the gout? O how often have we seen them, even immediately after they are anointed and thoroughly greased, till their faces glister like the keyhole of a powdering tub, their teeth dance like the jacks of a little pair of organs or virginals, when they are played upon, and that they foamed from their very throats . . .'

Shakespeare refers to 'the powdering tub of infamy' and it is possible that the nursery rhyme which begins 'Rub-a-dub dub, three men in a tub' has a similar connotation.

Whether mercury ever cured syphilis—or merely suppressed the symptoms—is an open question. In the 16th century, before the causative organism (the spirochete) had been domesticated by prolonged contact with man, the disease was more virulent than it is today. The early symptoms included severe ulceration of the skin, sometimes down to the bone. Mercurial treatment acted as cautery, burning away the ulcer—though sometimes the patient then suffered or died from mercury poisoning.

In later times, syphilis has become a less spectacular affliction. The early signs, including skin eruptions and fever, usually clear up, whether treated or not, in a few months; it is therefore not surprising that many remedies have been thought to be effective. An interesting clinical trial was made as long ago as 1812 by Dr William Fergusson. While serving as Inspector of Hospitals with the Portuguese troops who fought alongside Wellington's army, Fergusson observed that Portuguese soldiers received no treatment for syphilis, whereas their British comrades were usually treated with mercury. Over a period of two years, he found that the Portuguese soldiers recovered just as quickly as the British, and therefore concluded that mercury did not have any effect on the progress of the disease.

Though Fergusson shrewdly (if verbosely) observed

'. . . the virulence of the disease has become so much mitigated by reason of general and inadequately resisted diffusion, or other causes, that, after running a certain (commonly mild) course through the respective orders of parts, according to the known laws of its progress, it exhausts itself, and ceases spontaneously'

he did not know the whole story. It was not until a century later that general paralysis of the insane, locomotor ataxia and other grave disorders of the nervous system were recognised as the late manifestations of syphilis.

Fergusson's findings were ignored, but a more extensive trial was made many years later by Caesar Boeck, a Norwegian physician. During the period between 1891 and 1920, Boeck studied nearly 2000 patients with early syphilis, who were kept in hospital while the disease was in the infectious stage, but received no treatment. The outward signs of the disease disappeared, sometimes in a few weeks and sometimes after several months. Long-term follow-up studies were done, the first in 1929 and another between 1948 and 1951. These studies covered 80 percent of the original 2000 patients. They showed that about two-thirds of the patients 'went through life with a minimum of inconvenience despite no treatment for early syphilis'. This story is an example of the great respect given to mercury, by physicians and patients alike, without good evidence of any therapeutic benefits.

PIRATE AND PHYSICIAN

The mantle of Paracelsus fell on Thomas Dover (1662–1743), an English physician known to his contemporaries as the Quicksilver Doctor. He was also a part-time pirate. At a time when the British navy was reduced in strength, freelance privateers were licensed by the monarch to harry the French or the Spaniards and to take what they could find. Privateering had declined during the 17th century, but Queen Anne gave encouragement by the Prize Act of 1708, which provided that adventurers operating at their own expense would pay no royalty but would actually received a bounty, in addition to the value of whatever they captured. Dover made a large fortune in this way. During one of his voyages in the Caribbean, he rescued Alexander Selkirk, whose adventures are said to have inspired Daniel Defoe, the author of *Robinson Crusoe.*

Dover's Powder, a mixture of opium and ipecacuanha, is still well known, though no longer popular as a treatment for gout—which is perhaps just as well. The original recipe (in Dover's *magnum opus,* The Ancient Physician's Legacy to his Country) prescribed a heroic dose of

'forty to sixty or seventy grains in a glass of white wine Posset going to bed; covering up warm and drinking a quart or three pints of the Posset—Drink while sweating'.

This stupefying nightcap was reputed to ease the pain for a week or more.

Dover's book was a vehicle for his advocacy of the virtues of mercury. He prescribed the metal, in doses of a pound or more, for the treatment of asthma (where the effect, if any, was psychological) and intestinal obstruction, where the effect was gravitational. Reviewing the treatment of iliac passion (a complaint which embraced several abdominal emergencies, including that which now goes under the colourless name of appendicitis) he proclaimed:

> 'You need go no further for the cure of this fatal disease than take a pound, or a pound and a half, of crude mercury; and the late Queen Caroline had but taken the same remedy, I will avow she would have been well in twelve hours'.

The Queen died in 1727, after an operation to reduce a long-neglected hernia; Dover's plunger would not have done her any good.

Dover had many of the characteristics of Paracelsus, including a sharp tongue and a contempt for his fellow physicians. He was, however, brave enough to practice what he preached. There is no reason to doubt his account of his own experience, written in reply to an anonymous pamphlet attacking his methods:

> 'I have taken it myself above six and forty years, I have been in all sorts of Climates, and am now upwards of eighty and yet, I thank God, I enjoy a perfect State of Health'.

'I have taken it myself above six and forty years'.

These observations by Dover remind us of the great variation in tolerance to mercury. Some people appear to be able to absorb huge quantities without ill effect. Such indeed was the experience of Boerhaave (1668–1738), Professor of Medicine in Leiden. His interest in mercury arose from his respect for the alchemists. He did not spend his time in trying to make gold, but became adept at the distillation of mercury—a favourite activity of the alchemists—and distilled one sample more than 500 times. His devotion to this task was so conspicuous that he must have inhaled a great amount of mercury vapour, but he seems to have suffered no harm, nor could his colleagues find any evidence of relevant damage in the *post-mortem* examination which they conducted.

MERCURY IN THE NURSERY

Mercury continued to be widely used in medicine until well into the 20th century—not always under the supervision of doctors. Among the homely remedies recommended by Mrs Isabella Beeton, in her celebrated *Book of Household Management* (first published in 1861) were heroic concoctions of mercury (as well as lead and antimony) for the relief of almost every pediatric ailment. Measles, teething, diarrhoea, fits . . . all were to be treated with grey powder (a mixture of mercury and chalk) or calomel. Not surprisingly, the treatment was sometimes fatal.

A widely-used book of pharmaceutical formulas, published in 1929, offered more than 100 mercurial preparations, including 40 containing corrosive sublimate—in gargles, eye drops, mouth wash, soap, ointment and a variety of pills. Among the milder remedies was the Army's legendary Number 9 pill—containing calomel and rhubarb.

The difficulty experienced by even wise observers in recognising the hazards of mercury is well illustrated by the story of pink disease. This distressing and sometimes fatal illness, occurring in children up to about two years old, was for a long time a complete mystery. It was suggested in 1942 that the cause was mercury poisoning from teething powders containing calomel. This suggestion was supported by many pediatricians but drew opposition, sometimes ferocious, from other quarters. In 1952 the Minister of Health told the House of Commons that there was not enough evidence to justify any action—or even general publicity—about the mercury hazard to infants.

The issue was put beyond doubt by the concerted activities of three men in Stoke-on-Trent—the Medical Officer of Health, the Coroner and a pediatrician. The deaths of two infants in 1953 were reported to the

Coroner, who boldly returned verdicts of death from bronchial pneumonia, due to pink disease resulting from chronic mercurial poisoning from teething powders.

The Home Secretary refused, shortly afterward, to forbid the sale of mercurial teething powders; the manufacturers took a more responsible view and removed the product, under the familiar pretext of a new and improved formula. The Medical Officer of Health then confiscated all of the mercurial teething powders in pharmacies and other shops, replacing them by harmless preparations. The result of this enterprise was drastic. Up to 1953, the children's wards in Stoke (as in every other town) were hardly ever without cases of pink disease; after the withdrawal of the offending powders, not another case was seen in Stoke. Elsewhere, isolated cases continued to appear for a while, but by 1962 the death rate from pink disease in England and Wales, which had been about 50 per year until 1953, had fallen to zero.

MERCURY IN COSMETICS

Soaps and creams containing mercury compounds have been used for many years in African and Caribbean countries to lighten the skin. Mercury absorbed from such preparations was shown in 1972 (by studies including activation analysis of hair and nail samples) to be the cause of kidney disease, sometimes fatal, among young women in Kenya. The sale of soaps and cosmetics containing mercury has been forbidden in EEC countries since 1976, but manufacture for export to Third World countries continues. EEC legislation does not prevent these materials from being sent back to Europe for sale in immigrant communities. This legal loophole attracted attention in 1987 when unfavourable publicity provoked an English manufacturer of mercurial soap to move its factory from Lancashire to the Irish Republic, where further protests greeted its arrival.

CHAPTER 5

Our Daily Lead

Suetonius suggested apathy and gluttony. Other scholars have considered the influence of malaria, plague, over-cultivation of the soil, debauchery and climatic change. It is not difficult, after the event, to see some relevance in all of these speculations—and even in the proposition that the fall of the Roman Empire was brought about by an epidemic of lead poisoning.

Though the evidence is all circumstantial, it is beyond doubt that the Romans absorbed considerable amounts of lead. Having no cane sugar, they used sapa (contrentrated grape juice) for sweetening. The boiling down was done in lead pots; according to one contemporary writer, strips of lead were put into the juice to retard fermentation. Cookery books of the time advised that the use of brass or copper vessels gave the product a bitter or metallic taste. Recent experiments have shown that sapa, made in Roman style, has an agreeably sweet taste—and a very high lead content. Grape juice, like all fruit juices, contains organic acids capable of dissolving lead. Many of the salts formed in this way have a sweet taste; lead acetate, once widely used in medicine, was known as sugar of lead. Sapa was used in cookery, as well as in making wine and other beverages; wine appears in most of the recipes in the authoritative cookery book written by Apicius, who flourished in the first century. Aristocratic Romans were notoriously fond of wine and food. Recent estimates suggest that they absorbed 5-10 times as much lead as people living in Western countries today; plebeians and slaves had much smaller intakes, not enough to do any harm.

Lead poisoning was certainly known in ancient times, as an occupational hazard among workers engaged in the mining or smelting of the ores, and from experience with patients who had swallowed lead salts accidentally or as victims of homicide. But the Roman physicians were more concerned about the increasing prevalence of podagra, an affliction now known as gout. The causes of this disease were commonly identified as drunkenness, gluttony and general debauchery. The possibility of an association with lead poisoning was not perceived in Rome and, indeed, was not confirmed until 1859, when Sir Alfred Garrod reported that between a quarter and a third of the many patients whom he treated for gout were plumbers or painters and were suffering from lead poisoning. Similar findings have since been reported by many physicians. Some have suggested that the prevalence of gout among the British upper classes during the 18th and 19th centuries was a consequence of over-indulgence in wine contaminated by lead.

THE FALL OF ROME

The suggestion that lead poisoning was the major cause of the fall of the Roman Empire in the fifth century was first made by Gilfillan in 1965 and has since been developed by other medical historians. There is little doubt that the aristocracy declined in numbers and intellectual ability. Though the fall in the birth rate can be explained in many ways, the well-documented prevalence of sterility, miscarriages, stillbirths and infant mortality is consistent with a diagnosis of epidemic lead poisoning. But it is unlikely that this is the whole story, for the decline of a great civilisation is always a complex process.

The city of Rome, at its height, had a population of more than a million and was not in equilibrium with its environment. Agricultural productivity declined through over-cultivation of the land. As the bureaucracy expanded, it became less efficient. The far-flung army and colonial administration eventually absorbed more resources than they returned to the city. The Roman Empire outgrew its ability to adapt to political and environmental change.

THE POISONED CUP

Though the hazards of exposure to lead have been well known for more than 2000 years, they have not yet been effectively controlled. In 1960 a

British villager was admitted to hospital suffering from severe abdominal pain. He went home after treatment, but before long returned in an even worse state. Investigations showed that he had been poisoned by elderberry wine contaminated with lead. In 1967 an American physician, complaining of headache and exhaustion, was found to have liver damage and abnormal red blood cells. He had been poisoned by lead in a well-known carbonated beverage, which he drank every evening from a home-made jug. A middle-aged Englishman, suffering from severe abdominal pain, was found to have been poisoned by lead present in home-made cider which he had enjoyed for 30 years.

In the first two of these cases, the illness had the same cause—absorption of lead from glazed pottery. This is a hazard which has long been recognised. As long ago as 1774, Dr Rice Charlton of Bath wrote of six patients:

'who became paralytic after drinking cyder brought to them at harvest work, in a new earthen pitcher, the inside of which was glazed'.

Earthenware vessels are porous and therefore unsuitable for the storage of liquids until they have been waterproofed, usually by glazing. A glaze is a thin film of a glass-like material, produced when the vessel, painted with a suspension in water of silica and other materials, is heated in a kiln or oven. Silica melts at an inconveniently high temperature; glazes containing oxide or carbonate of lead have a much lower melting point and other useful properties, but are partly dissolved on contact with acid. The death in 1970 of a Canadian child was due to lead poisoning, from apple juice kept in an earthenware jug. Only in this century has progress been made in the development of glazes containing lead silicate, which are relatively insoluble, or lead-free glazes. Safety standards which have been imposed by legislation in Britain and some other countries are not observed everywhere else. Soluble glazes are regularly found in pottery made by primitive methods in many Middle Eastern countries for use in homes and restaurants. There the possibility of excessive intake of lead is enhanced by the liberal use of lemon juice and vinegar in traditional recipes.

A FAMOUS COLIC

The detailed study of lead poisoning began more than three centuries ago. In 1572 the peasants of Poitou in France were prostrated by the most celebrated stomach-ache in history. This affliction was later dignified by the name of colica Pictonum, from the Pictones, a Celtic tribe (sometimes identified with the Picts) who lived in Poitou.

In 1656 Samuel Stockhausen, works doctor at the lead mines of Goslar in Germany, declared that epidemics of colic, which were not uncommon, were caused by exposure to lead. Soon afterwards, contamination of wine by lead was recognised as a common cause of the affliction; lead oxide (litharge) was often used to sweeten sour wine—in which it was converted to lead acetate, the aptly-named 'sugar of lead'. Devonshire colic was a serious health problem in the 18th century, among villagers who made or drank cider. Sir George Baker, the eminent English physician was much criticised in Devonshire (his native county) when he uncovered the cause. The juice from the apple presses was commonly collected and stored in lead vats. In Herefordshire, where vessels of wood, stone or iron were used, colic was almost unknown.

Josiah Wedgwood, the celebrated English potter, was disturbed when he learnt of the danger lurking in glazed vessels. He set out to make a glaze free of lead, but the traditional material continued to be used and lead poisoning remained a serious occupational hazard in his industry. Josiah Wedgwood II showed less compassion than his father when he gave evidence to a parliamentary committee in 1816. Asked whether any part of his business was unwholesome, he replied:

> 'There is a part of the business that is unwholesome; it is that part of the business that is connected with the applying of glaze upon the surface of the ware . . . that glaze is composed in part of white lead, and like other businesses in which workmen have to do with lead, they are, if careless in their method of living, and dirty, very subject to disease'.

Until stringent legislation was enacted early in the present century, little was done to protect pottery workers from lead poisoning. It was, however, a tradition that lead smelters, painters and pottery workers were provided by their employers with a daily glass of milk. The belief in milk as an antidote to lead is very old, though of doubtful validity. It has the authority of Ramazzini, who advised in 1700 that lead miners should gargle with milk and wrote, a few years later:

> 'To correct the desiccation incurred by the lungs from breathing air saturated with metallic vapours . . . I recommend whey and dishes made with milk'.

The custom achieved legal force in Britain in the Regulations for the Manufacture and Decoration of Pottery, issued in 1913. Experiments to assess its usefulness have given conflicting results. Other measures, enacted in the face of much opposition from manufacturers, have been more effective. During the first 40 years of this century there were, on average, about seven deaths per year from lead poisoning among pottery workers in North Staffordshire. After 1953 there were none.

Another source of lead poisoning was described in 1745 by Thomas Cadwalader, an American physician, in his essay (printed and published in Philadelphia by Benjamin Franklin) on the West Indian Dry Gripes, an affliction caused by drinking rum distilled in lead vats. This method of manufacture was forbidden by law in the Massachusetts Bay Colony as early as 1723. Lead poisoning persists to this day among the moonshiners of the southern United States and their customers. Illicit stills, which may have to be abandoned in haste, are usually made from makeshift materials, including piping with soldered joints and old car radiators. The whisky that they deliver is often heavily contaminated by lead, which offers the revenue authorities a simple way of distinguishing the product from genuine duty-paid liquor.

DEATH OF AN EXPEDITION

Lead poisoning has recently been invoked to explain the disappearance of the expedition, led by the eminent Arctic explorer Sir John Franklin, which sailed from the Thames in May, 1845 with instructions to find the Northwest Passage, which would provide a short route between North America and the Far East. The expedition was lavishly provisioned with food and liquor. The two ships were fitted with steam heating, desalination equipment, scientific apparatus—and mechanical organs which played ten hymn tunes. They were last seen in Baffin Bay in July, 1845. As time passed, fears mounted for their safety and fresh expeditions (as many as 40 by 1860) were sent to look for them. The fate of Franklin's party was the inspiration for decades of Arctic exploration. Three graves were found in 1850; by then it was apparent that Franklin and his 129 shipmates had died, leaving only a few traces. It was generally believed that they had been overtaken by scurvy and starvation.

In 1981 a scientific party led by Professor Owen Beattie, a young Canadian anthropologist, found bones of three Eskimos and one member of the Franklin expedition. The Eskimo bones contained lead at levels between 22 and 36 PPM but one sample from the Franklin crewman showed 228 PPM. Lead has on rare occasions been found at concentrations above 200 PPM in rib bone from healthy inhabitants of Glasgow—who had, of course, been drinking water with high lead content throughout their lives. A lead content of 228 PPM achieved during the expedition—lasting for only a year or two—would probably have been accompanied by symptoms of poisoning. Professor Beattie made a tentative diagnosis of lead poisoning, acknowledging that bone lead reflects long-term exposure rather than recent intake, and that the British environment in which the sailor had spent most of his life was quite heavily contaminated by lead.

Seeking further evidence, Professor Beattie returned to the Arctic in 1984 and exhumed the bodies of three of Franklin's crew. Autopsies suggested that all three had suffered from tuberculosis and had died from pneumonia. They may, however, have been weakened by lead poisoning. Hair samples showed lead at levels between 138 and 657 PPM. Hair, growing at about a centimetre a month, reflects recent exposure to lead. Concentrations of a few hundred PPM have been found in people living near smelters, but comparable levels found in the Arctic must have a different cause.

The expedition was provided with a large amount of canned meat. From examination of empty tins among the remains, Professor Beattie confirmed earlier suspicion that much of the meat was putrid, because of defective workmanship in the soldered joints. Scurvy, cold, tuberculosis, pneumonia and starvation may all have been implicated in the deaths of Franklin and his comrades—but lead poisoning probably contributed to the catastrophe.

LEAD AND MENTAL ABILITY

Although lead poisoning of the severity and frequency encountered in earlier centuries is now uncommon, there is increasing concern about the consequences of low level exposure of large populations. Anxiety is generated particularly by claims that the behaviour and mental development of children may be impaired by exposure to lead at levels which are commonly present in the environment.

In one type of experiment, the case-control method is used to study a group of mentally retarded children and a control group of normal children—matched, as far as possible, for age, sex, social class and other relevant factors. In some investigations, blood lead levels in the two groups have been studied; in others, the lead levels in drinking water in the children's homes have been compared. Surveys of these kinds have usually suggested an association between mental retardation and exposure to lead in the environment, as measured by the concentration in blood or drinking water. The findings have been criticised on several grounds. Blood lead reflects recent exposure, so a single sample may not give a reliable indication of long-term exposure. Lead levels in drinking water must also be interpreted cautiously; a child's fluid intake is not taken entirely from the tap. There have also been difficulties in matching the control and study groups for all of the relevant variables; allowance has not always been made for the influence of birth order in the family and of maternal age, both of which are known to affect the incidence of mental retardation.

A different approach was taken in a study made in the vicinity of a lead smelting works in London. Blood lead levels were higher in children living within 400 metres of the works than in those living further away. But tests of intelligence and behaviour showed no difference between children with high levels of lead in blood and those with lower levels. Children who had spent the first two years of their lives more than 500 meters from the smelter were less intelligent and more disturbed than those who had lived in the more heavily contaminated region nearer to the works.

Another study compared the eleven-plus school examination marks for a large group of children living in a polluted area near a battery factory in Birmingham and other groups living in areas with little or no environmental pollution by lead. The examination marks were higher among the children living in the more highly polluted area.

In the Shoshone Lead Health Project, conducted in Shoshone County, Idaho, during 1974 and 1975, a variety of tests were made on children living at various distances from a large lead mining and smelting complex. Although many of those living within a mile of the smelter showed blood lead levels well above µg per 100 millilitres—the level widely regarded as the threshold for damage to health—the conclusion of the survey was that no mental or physical health defects could be attributed to lead pollution.

After reviewing evidence of the kind produced by these and other investigations, the Royal Commission on Environmental Pollution advised in 1983 that it was not able to establish whether the observed differences in mental development and behaviour were due to toxic effects of lead, to confounding factors or to both. The confounding factors include parental IQ and social, educational, dietary and hygienic conditions in areas of urban deprivation where the environment is abnormally polluted by lead.

SOURCES OF LEAD

Though the original charge against lead—that it restricts the mental development of children—has not been proved, the average level of lead in blood in the British population is about a quarter of the level at which unmistakable evidence of harmful effects has been reported. Since the safety margin here is much less than would be thought acceptable for other toxic substances, it is clearly desirable that the exposure of the population to lead should be reduced. To see how this might be done, it is necessary to look at the major sources of exposure to lead.

(1) Food normally constitutes the largest intake, at about 100 μg per day for adults and half as much for two-year old children. It is important to distinguish between *intake,* which is what goes into the mouth or lungs, and *uptake,* which is the proportion absorbed from the gut into the blood stream. Uptake of lead is about 10 percent in adults and 50 percent in two-year-olds.

(2) Water normally accounts for less than 10 percent of the lead intake; where tap water is heavily contaminated by lead, this proportion may rise to 60 percent.

(3) Petrol has, for about 60 years, contained tetra-ethyl lead and tetra-methyl lead, added to allow engines to run with higher compression ratios and therefore better fuel economy. Of the lead absorbed from the lungs and gut, petrol accounts for about 16 percent in rural or suburban areas and 30 percent or more in inner cities.

(4) Alcoholic drinks make an uncertain contribution. There is evidence that blood lead level increases with increasing alcohol consumption. Although the concentration of lead in British beer is little more than in average tap water, the consumption of four pints per day increases the uptake of lead by about 20 percent.

(5) Tobacco contains a small amount of lead, from pesticide residues. Smoking 40 cigarettes a day increases the uptake of lead by about 20 percent.

(6) Dust contains an appreciable amount of lead—less than 0.1 percent indoors but up to 1 percent or more in busy inner city streets. A two-year-old child ingests about 100 mg of dust and soil per day; this accounts for more than half of his intake and uptake of lead.

(7) Cosmetics and traditional medicines. In many Asian and African countries (and in immigrant communities in Britain) lead and other heavy metals are absorbed from cosmetics and traditional medicines. Surma is a material believed to strengthen children's eyes, improve their

appearance and protect then against disease. In Britain, it is sometimes made by hakims (traditional healers) in Asian communities and sometimes imported from India and Pakistan. It is applied directly to the surface of the eye from an ornamental applicator. An extensive study of Surma, made by a team of scientists at Nottingham University, showed that some varieties were composed largely of lead sulphide. Children who had been exposed to Surma were found to have abnormally high concentrations of lead in the blood; a 4-year old boy was admitted to hospital suffering from lead poisoning.

Kushtay (plural of Kushta) are preparations, made in Britain or imported, and prescribed by hakims as tonics or aphrodisiacs. Of 37 samples analysed by the Nottingham team, 11 contained mercury, lead or arsenic as major constituents. Several other Asian medicines were found to be of relatively innocuous composition, though some contained undesirable amounts of toxic metals. Tiro, a material similar in composition and reputation to Surma, is widely used in Nigeria for the treatment of eye diseases and as an eye cosmetic. A study in Kuwait, reported in 1981, found that many varieties of Kohl, used as an eye shadow by adults and often applied also to children's eyes, contained lead at concentrations up to 92 percent. Lead poisoning was diagnosed in 24 infants, of whom four died and six suffered long-term damage. In four cases it was found that sick infants were exposed to fumes produced by heating lead or lead sulphide—a traditional practice known as Bokhoor.

REDUCING LEAD INTAKE

There are obviously many ways in which lead intake might be reduced. Unfortunately they do not all lead to a corresponding reduction in uptake. At high levels of intake, blood levels change with the cube root of intake; in other words, an eightfold reduction in intake produces only a halving of uptake. This relationship possibly reflects a physiological defence mechanism to limit the harmful consequences of high intake—but it also indicates that apparently substantial reductions in intake will not have much effect on uptake. These considerations do not, of course, undermine the case for reducing environmental lead by all reasonable means. The prospects of successful action can be assessed in a number of ways.

1. *Food.* The lead content of food crops comes both from the soil and from the air. Antarctic soil contains about 10 PPM of lead, which may be taken

as the natural level in the absence of man-made contamination. Urban soil in Britain and other countries now contains lead at levels of several hundred PPM; in rural areas, concentrations of 100 PPM are quite common. It is probable that most of the lead in contemporary soil (and therefore in food) comes from airborne contamination during the past century, first through increasing consumption of coal and, more recently, through increasing use of leaded petrol—a trend which is now diminishing. Smaller contributions to the lead in soil come from the use of sewage sludge as a fertiliser and from various industrial processes. Even if man-made pollution ceased entirely, the concentration of lead in the soil would not return to its pristine level for centuries. For this reason, dietary lead is not likely to diminish substantially in the near future.

2. *Petrol.* The move towards unleaded petrol began in the late 1960s in California, where clean air legislation (motivated by objections to the prevalent smog in urban areas) encouraged the fitting of catalytic converters in motor vehicle exhaust systems. These devices prevent the emission of unburnt fuel, carbon monoxide and oxides in nitrogen, but their operation is inhibited by lead. During the past ten years, progressive reduction in the lead content of petrol, now motivated by more direct considerations of health, has been achieved in Britain and other countries. The concentration of lead in petrol in Britain, in grams per litre, was about 0.84 in 1971, 0.4 in 1983 and 0.15 in 1985. Complete elimination of lead from petrol would reduce the daily uptake of the average adult by a few micrograms, but would be more significant in young children. The contribution of petrol to the lead content of dust is not known, but may be substantial.

3. *Cosmetics and traditional medicines.* Legislative action taken in Britain to forbid the importation or supply of preparations of this kind which are likely to be harmful to health will not be effective, since the materials are easily introduced in gift parcels or personal luggage. Health education among the vulnerable communities will be difficult in the face of traditions which have lasted for centuries.

4. *Water.* The average concentration of lead in tap water in most British houses is well below the limit of 50 µg per litre recommended by the World Health Organisation. Hard water, which is common in England, is slightly alkaline and contains salts of calcium which form a protective scale, preventing the uptake of lead from piping. In Scotland, where the water is soft, slightly acid and, in older houses, delivered to the tap

through lead pipes from lead storage tanks, much higher levels of lead are sometimes found. In 1972, members of four families living in rural Scotland were treated in hospital for lead poisoning. Their water supplies came from rivers or springs with normal lead levels, which were greatly increased (to over 3000 μg per litre) after storage in lead tanks and passage through lead pipes. The patients improved in health after the lead plumbing in their homes was removed.

General replacement of lead plumbing (by copper or plastic tanks and piping) would be desirable, and will be achieved in time. Meanwhile, simpler measures can be very effective. In 1977 the median lead level in Glasgow tap water was 120 μg per litre. Steps were taken to make the public water supply slightly alkaline (by the addition of calcium hydroxide to the reservoirs) and to reduce the solubility of lead in the water, by the addition of sodium dihydrogen orthophosphate. By 1980, the median lead level was down to 13 μg per litre. This spectacular improvement, made without publicity in a short time and at little cost, was a major contribution to preventive medicine.

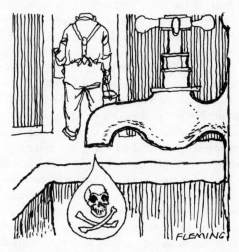

General replacement of lead plumbing would be desirable.

DIRTY WATER

By contrast, a recent controversy involving the Department of the Environment and the Water Research Centre illustrates difficulties that can frustrate efforts to reduce the contamination of drinking water by lead. Soon after the opening, in 1971, of a new hospital in Glasgow, complaints arose from nursing and medical staff concerning grease, fragments of metal and dark brown coloration in both hot and cold water. After inconclusive discussions extending over more than a year the Health Board's environmental scientists were asked to investigate the situation. Although the hospital's plumbing system was constructed of copper tubing, the water was found to contain lead at levels up to 4000 μg per litre as well as copper at levels up to 18 000 μg per litre. Considerable amounts of grease and fragments of lead and rusty iron were also found in the water. Since the hospital had used more than 50 000 tons of water by the time of these findings, it was apparent that the contamination could not be attributed to the fortuitous debris said to be found in every new piping system.

The cause of the trouble was found by visual inspection, reinforced by elementary chemical considerations. The hot and cold water systems were made from thin-walled copper tubing which, though relatively cheap, cannot easily be bent or threaded. Consequently the use of this material (instead of the more common thick-walled copper tubing) requires a greater number of joints and does not allow the use of compression fittings, based on threaded sleeves. It is necessary instead to make capillary joints. In this technique a copper sleeve, with a built-in ring of lead solder, is slipped over the ends of the tubes to be joined. Gentle heating with a blowlamp releases the solder, which forms a thin film and makes a watertight joint of adequate mechanical strength. This procedure required moderately skilled workmanship.

In the Glasgow hospital, the urge to save time was manifested by the application of liberal amounts of soldering flux to the pipe ends, and of additional solder to the outsides of the joints. The flux was a witches' brew of zinc and ammonium chlorides in a base containing tallow, lard oil, olive oil, lanolin and wool wax. This flux was certainly effective in promoting the flow of solder, but unfortunate consequences followed the slow removal into the flowing water of the soluble chlorides. The presence of dissimilar metals in a slightly acid solution of ammonium and zinc chlorides made a favourable environment for a variety of electrolytic actions, involving copper and lead, as well as iron and steel from thermostats at various places in the system. The plumbing system was eventually replaced, at great cost. The same difficulties occurred in a number of other hospitals that were built later. Defects in the plumbing system and other parts of the Glasgow hospital were referred to arbitration. Legal arguments continued for many years but were ended in 1987 when the main contractors paid £3.75 m in settlement.

After these misfortunes came to light, the Department of the Environment commissioned the Water Research Centre to investigate the hazard of lead pollution through electrolytic action in thin-walled copper plumbing systems. Their report, published in 1981, came to the same conclusion as the earlier Glasgow study and advised that lead-based solder should not be used in pipes carrying drinking water. The Department of the Environment did not accept this recommendation, though they did urge local authorities to avoid the problem by raising the standard of workmanship.

In its later, as well as its earlier, stages the lead story illustrates the anxieties and difficulties attending the control of many environmental hazards. The evidence of damage to health, even in places where pollution levels are high, is not always convincing. Actions by national and local authorities to reduce exposure have been considerably influenced by public opinion. Pressure groups and concerned individuals do not need to demonstrate harmful effects. They argue, reasonably enough, that lead does no one any good and that even a small possibility of damage to health is enough to justify its elimination. We are surrounded by potential poisons and we cannot avoid them completely. But it is likely that other trace elements, now waiting in the wings, will be brought forward as targets for public concern and indignation in the future.

CHAPTER 6
Death in the Pot

The lofty dining room of St Salvator's Hall, a student residence in the ancient university town of St Andrews in north-east Scotland, was well filled on 15 January 1945 as more than 100 young men sat down to lunch. The main dish was sausages, supplied by a long-established and highly reputable firm of butchers in the town. Before the end of the meal, many students were feeling ill. Ninety of them had stomach pains, sometimes followed by diarrhoea and vomiting. Later in the day some 60 other cases of acute illness occurred throughout the town.

Prompt investigation by health authorities showed that only those who had eaten the sausages were affected. Among them were a man of 52 and a woman of 48 who both died two days later. A bacteriologist who examined samples of unsold sausages found nothing unusual, but chemical analysis indicated that the first batch produced by the sausage-making machine had contained a large amount of arsenic trioxide. This batch had been sold to members of the public. Smaller, though still substantial, amounts of arsenic were present in the second and third batches, which had been supplied to the students' residence. Calculation suggested that about 250 grams of arsenic trioxide had been added to the sausage meat before processing began on the morning of 15 January; the second and third batches were presumably contaminated by residues remaining in the machine.

The circumstances surrounding this outbreak are still mysterious. No other trace of arsenic was found in the butcher's shop. Stocks in local laboratories were intact. Records of retail and wholesale pharmacies in the neighbourhood showed no unusual purchases. It was however found that a rat catcher living in an adjoining county had bought 80 lb of arsenic trioxide during the previous two years. Whether—or how—any of this material found its way into the butcher's shop is not known. It was suggested, some 40 years after the event, that German agents had poisoned the sausage meat with the intention of killing a number of airmen who were attending a navigation course in the university; this explanation seems improbable.

One of those who ate the poisoned sausages was James Renwick, then a medical student and later a university teacher in Glasgow and in London. In 1981 he reported on a follow-up study undertaken to search for long-term effects of the poisoning. Information was obtained from 62 survivors. One had been plagued with rodent ulcers of the skin since 1951 and two reported having had single rodent ulcers. All three had spent time in the tropics. It was considered that ultra-violet radiation was the primary cause. It is well known that prolonged exposure to arsenic enhances the susceptibility of the skin to damage by ultra-violet radiation. Dr Renwick's study suggests a small possibility that a single substantial dose might have a similar effect.

A VERSATILE ELEMENT

The mystery of the poisoned sausages is typical of the many fascinating stories—some true and some legendary—which have emerged during the long association of arsenic with man. Arsenic compounds have been widely used both as medicines and as poisons. They have been suspected both of causing cancer and of retarding the growth of tumours. In many animals—and perhaps in man—arsenic is essential for healthy life. More recently arsenic has had a decisive role in the development of semi-conductor materials used in transistors, silicon chips and computers.

Arsenic is widely distributed in the environment, being present more abundantly than most elements in rocks, soil and sea water. It often occurs in association with iron, copper, lead or silver. Sulphides of arsenic are present in the earth and were used in early times both as pigments and as medicines. The internal environment acquired arsenic also through the growth of the metallurgical industries. Copper objects dating from 3000 BC contain concentrations of arsenic as high as 12 percent, reflecting the presence of the element in the crude copper ores. Many copper objects made about 2000 BC were quite pure; the change is attributable to improvements in smelting techniques.

Used in early times both as pigments and as medicines.

Arsenic is obtained commercially as a by-product from the smelting of ores of copper, zinc and lead, in which it is present, sometimes to the extent of 15 percent, as an unwelcome impurity. When these ores are roasted, arsenic is released as the trioxide and accumulates in the flue dust, which is then removed and heated gently. Arsenic trioxide sublimes (that is, changes from solid to vapour without melting) and passes into cooling ducts, where it condenses and may be recovered as a white solid. Arsenic trioxide (sometimes known as white arsenic), an almost tasteless and often effective poison, is the substance colloquially known as arsenic. By careful heating it can be reduced to the element itself, which is a grey solid with a metallic sheen—though it is not a true metal in the chemical sense.

The industrial demand for arsenic has increased greatly in recent times. A hundred years ago the total world production of arsenic trioxide was about 10 000 tons. At that time, arsenic compounds were used in the manufacture of lead shot and glass, as well as in medicine. Copper hydrogen arsenite, generally known as Scheele's green, was widely used as a pigment in wallpaper and textile printing and even in confectionery. By the early part of the 20th century, arsenic compounds were being used in wood preservatives, sheep dips, fly papers and a variety of agricultural pesticides. Today the annual consumption of arsenic is about 50 000 tons; a further quantity of at least 10 000 tons is added to the environment each year through the burning of coal.

As arsenic is so widely distributed through the earth and sea, its presence in almost every foodstuff is not surprising. In Britain the average dietary intake is about 50 µg per day. About 70 percent of this comes from fish; meat provides about 20 percent, nearly all from chicken and pork, with hardly any from beef. The reason for this difference is that arsenic compounds are widely used for improving growth and controlling disease in pigs and poultry. These compounds were originally used to treat coccidiosis, a wasting disease in chickens; the growth-promoting effect was discovered accidentally. The increase in weight is only about 3 percent, but is more than enough to cover the cost of the drug. A bottle of wine may contain as much as 50 µg of arsenic, from pesticides used in vineyards.

Seaweed (or kelp) tablets, sold in health food shops, contain significant amounts of arsenic. The dose regime recommended by the manufacturers of some of these products can raise an individual's daily arsenic intake by 50 µg, though this increase would not be likely to do much harm. A person who eats fish every day may take in as much as 250 µg of arsenic per day—but even this amount is unlikely to be dangerous; doses of as much as a few mg were often prescribed in earlier centuries without causing any apparent ill effects.

A DANGEROUS DRINK

A serious outbreak of arsenic poisoning occurred in north-west England in 1900. Writing in the *British Medical Journal* on 24 November of that year, Dr Ernest Reynolds drew attention to a recent increase in the number of cases of alcoholic neuritis admitted to hospitals in Manchester and neighbouring towns. Peripheral neuritis was a well-recognised condition, characterised by paralysis and wasting of muscles in the arms and legs, and by loss of function in sensory nerves. When these symptoms occurred in heavy drinkers, they were attributed to alcoholic neuritis, although the connection had not been firmly established. One physician who expressed polite scepticism was Sir William Gairdner, Professor of Medicine in Glasgow. When he was examining in Manchester in the 1880s, he was surprised by the large number of patients with peripheral neuritis—an affliction that he had seen only twice in his whole career in

Glasgow, where the consumption of alcohol was at least as great as in Manchester.

Dr Reynolds was the Visiting Medical Officer to the Manchester Workhouse Hospital. During one ward round in November, 1900, he counted 25 cases of alcoholic neuritis. While pondering on the possible cause of this outbreak, he was reminded by the Resident Physician that many of the victims were also suffering from shingles. That was the clue that he had been seeking. He believed that shingles was sometimes a manifestation of arsenic poisoning; therefore the patients must have been poisoned by arsenic in their beer. Dr Reynolds did not, of course, know that shingles *(herpes zoster)* is a virus infection—but there is some similarity with a skin condition resulting from arsenic poisoning.

On 17 November Dr Reynolds analysed a sample of the beer that some of the patients had been drinking. It contained an appreciable amount of arsenic. On 20 November, Dr Sheridan Delepine, Professor of Pathology at Owens College (now Manchester University) reported similar findings from examination of beer from 14 local breweries. On the following day, when samples of various materials were obtained from five breweries, it was found that glucose supplied by one manufacturer (Bostock and Company) was contaminated with arsenic. On 24 November the source of the arsenic was traced to sulphuric acid made by Nicholson and Son of Leeds. This company had, until March 1900, supplied reasonably pure sulphuric acid—but had then changed to a cruder product, made by the lead chamber process using pyrites (iron sulphide) in which arsenic had long been a known impurity.

After Dr Reynolds reported his preliminary findings in the British Medical Journal, many more cases came to light and it was apparent that an epidemic was in progress, not only in Manchester but also in other towns in Lancashire and the adjoining counties. The wheels of government moved quickly in those days. On 4 February, 1901—less than two weeks after his accession to the throne—King Edward VII signed the document appointing a Royal Commission, under the chairmanship of the venerable Lord Kelvin, who was then 76 years old; he had retired in 1899 after serving for 53 years as Professor of Natural Philosophy in the University of Glasgow. The commission found that more than 6000 people had been poisoned, with at least 70 deaths. All of them had drunk beer made from materials including glucose and other sugars supplied by Bostock and Company. These sugars, which have an important part in the brewing process, are made by the action of sulphuric acid on starch.

The report of the Royal Commission was a remarkable document, demonstrating clarity of thought, scientific insight and appreciation of political and administrative factors in the management of environmental hazards. Advising that 'it should be the aim of the manufacturer to exclude arsenic altogether', the report acknowledged that it was unrealistic to suppose that beer or food could ever be completely free of arsenic. The Commission recommended the prescription, with legal force, of 'a proportion of arsenic to be regarded as altogether inadmissible'. This conclusion was of historic importance, for it was the first suggestion ever made that toxic hazards in the environment should be controlled by prescribing maximum permissible levels. The figure proposed for beer 'or other liquid foods' was 0.01 grain of arsenic per gallon. For solid foodstuffs, the figure was 0.01 grain per pound—a level not much different from 1 PPM, which was eventually written into legislation in Britain and other countries.

The justification for these recommendations rested on three principles, expressed by the Commission as follows:

(a) The recommended levels could be achieved without undue difficulty by good manufacturing practice.
(b) the presence of arsenic at the recommended levels could be tested by existing analytical techniques.
(c) the presence of arsenic in foodstuffs at the recommended levels did no harm to rats.

These three criteria are essentially the same as those used today in defining maximum permissible doses for radiation and many other occupational or environmental hazards.

The prophetic wisdom of the Commission did not stop here. They wrote:

'The fact of arsenic being excreted by the hair appeared to us to be of special interest. If it were established that persons taking comparatively small quanities of arsenic habitually excreted the poison in their hair to an extent which is readily appreciable by chemical tests, examination of hair might be of much value in cases where it is important to obtain indication of the past history of a patient in regard of arsenic'.

STRONG MEDICINE

In early times arsenic was used chiefly as the sulphide (realgar) or the trisulphide, usually known as orpiment—originally auripigmentum, from its golden colour. These materials, made into ointments or plasters, were prescribed for the treatment of cancer, skin disorders and eye diseases, or used in amulets against the plague. The popularity of arsenic among doctors and patients grew rapidly after the discovery, early in the 18th century, that the trioxide became more soluble when boiled with an alkali. Fowler's solution, a mixture of potassium arsenite and potassium carbonate prepared in this way, was widely used until quite recently and has not yet disappeared from clinical practice. During the 19th century, arsenic was recognised as a poison, but was recommended by eminent authorities for the treatment of malaria, asthma, tuberculosis, cancer, rheumatism, diabetes, bronchitis, epilepsy, headache, leprosy and alcoholic hangover.

was recommended by eminent authorities

A more rational therapeutic approach was pioneered by Paul Ehrlich, a German chemist who, early in the 20th century, began to search for the magic bullet'—a drug which would destroy bacteria circulating in the blood without killing or seriously damaging the patient. He concentrated his search on organic compounds of arsenic, which he invented by the hundred. Number 606, produced in 1909, was not successful against the infection for which it was intended, but turned out to be very effective in the treatment of syphilis. Afterwards named salvarsan, it was much used in this connection for many years, until superseded by penicillin. Salvarsan was the first man-made chemical to be effective against a major disease; its appearance marked the beginning of modern chemotherapy.

Many notorious poisoners relied on arsenic.

ARSENIC AS A POISON

The toxic hazards of the sulphides of arsenic must have been observed by the early physicians and their patients. Arsenic trioxide, said to have been discovered in the 8th century by Geber, a celebrated alchemist, was soon recognised as a poison. Though there were attempts in later centuries to restrict the sale of arsenic compounds, poisoners had many successes. The first reliable account of the use of arsenic for homicidal purposes appears in the record of the trial of Wondreton, a minstrel, who was charged in Paris in 1314 with the attempted murder of Charles VI, King of France, and other royal personages. His instructions, allegedly given by Charles the Bad, King of Navarre (brother of Charles VI) were to buy arsenic trioxide from apothecaries in various cities and to sprinkle it into soup and other food—where it would not be noticed, since it is almost tasteless. These efforts did not succeed, but there is little doubt that many notorious poisoners, including Catherine de Medici and the Borgias, relied on arsenic, which was often referred to as the succession powder.

Arsenic ... often referred to as Succession Powder

Poisoners often escaped retribution, because there were no reliable methods for detecting small amounts of arsenic. Sensitive techniques were developed during the 19th century, but justice was sometimes frustrated by the claim that the deceased had been in the habit of taking arsenic for medicinal or other purposes. This defence (p 118) was made more plausible by the prevailing belief that arsenic was useful as an aphrodisiac, or for improving the complexion.

THE ARSENIC EATERS

One of the most persistent—and preposterous—legends relating to trace elements tells that peasants living in the Austrian province of Styria regularly consume large quantities of arsenic without ill effects; indeed, the practice is said to improve the complexion, stamina and general well-being. If the habit is interrupted (the story goes on) the symptoms of arsenic poisoning appear and may progress to a fatal outcome.

The origin of this legend is well documented. It began with a paper by JJ von Tschudi in an Austrian medical journal in 1851. He explained that arsenic trioxide was bought from pedlars and taken in quantities of as much as 250 mg (normally a lethal dose) a few times per week. He went on to report:

> 'Poison eaters have a two-fold object in their hazardous indulgence. On the one hand, they look to obtain a fresh, healthy aspect and a certain degree of obesity. It is hence very frequently taken by peasants, youths and girls, in order to promote reciprocal liking, and it is, in fact, remarkable with what favourable results their views are attended. The juvenile poison eater shortly exhibits a very blooming complexion and a strikingly handsome exterior . . . The second object which the poison eaters have in view is to render themselves, as they express it, more airy or lighter—to facilitate respiration in ascending mountains'.

Arsenic eaters found, according to von Tschudi, that it was desirable to increase the dose around the time of the new moon. He claimed also that the feeding of substantial doses of arsenic to horses rendered them fat and plump, gave a bright glossy skin and greatly improved their general health. It was important that the arsenic should be given only at the time of the new moon. Von Tschudi, who travelled widely, spoke also of the practice, said to be common in South America and in parts of Europe, of eating corrosive sublimate. This is mercury bichloride, a most unpleasant poison. Two grams is a fatal dose but, according to von Tschudi, the British Ambassador to Turkey reported one case in which a quantity of 2.5 grams was swallowed every day without ill effect.

Von Tschudi's paper was reported in Chambers's Journal in 1851 and was given wide publicity also in James Johnston's book: *The Chemistry of Common Life,* first published in 1853. From that source, and from subsequent articles in Chambers's Journal, the story has been copied and improved by a succession of credulous writers. As with flying saucers and the Loch Ness Monster, the legend when once launched has attracted many supporters—even eye witnesses. In some versions the peasants are said to find arsenic among the rocks, eating it *al fresco* or taking it home to spread on their bread. Some commentators identify the material consumed as arsenic trioxide, but others favour arsenic element.

In a series of articles published in the *Association Medical Journal* (forerunner of the *British Medical Journal*) in 1856, Mr W B Kesteven, a London surgeon, demolished the legend, showing that statements made by von Tschudi were based on hearsay without any trustworthy

to render
themselves
more airy
or lighter

foundation. If the Styrian peasants did swallow some white powder, it was more probably harmless zinc oxide. Yet knowledgeable and reputable writers, even in recent times, have accepted the reality of arsenic and have sought in ingenious ways to explain why doses of poisonous material which would be fatal in any other part of the world are consumed with impunity by Austrian peasants—and why, contrary to the experience of every toxicologist (and to the dictates of common sense) the symptoms of arsenic poisoning among the addicts appear only when they stop taking the toxic substance.

According to von Tschudi, an alleged murderer, brought to trial in Austria in 1850, was acquitted because the court was persuaded that the victim had been an arsenic eater. This was perhaps the first use of the legal device which came to be known as the Styrian Defence. It appeared in slightly different form in 1857, and secured the release of Madeleine Smith, a young lady of respectable family who was tried in Edinburgh for the murder of her lover, Emile L'Angelier. There was no dispute that L'Angelier had died of arsenic poisoning, that Miss Smith had bought a substantial amount of arsenic not long beforehand and that she had the opportunity to administer the poison in cocoa. Her defence was that she had been told of the Styrian arsenic eaters and had used the arsenic as a cosmetic—applying it, mixed with water, to her hands, face and neck on the grounds that it was good for the complexion.

At that time, arsenic was sold only after being coloured with soot or indigo, which would not prevent its use as rat poison (the purpose for which it could legally be bought) but would make its administration to human victims more difficult. The arsenic which Miss Smith bought was mixed with indigo and would certainly not have improved her complexion. Defence counsel claimed that L'Angelier had, in 1852, spoken of the use of arsenic as a horse medicine, and had admitted taking some himself. It was suggested that he had committed suicide by taking a large dose of arsenic, though there was no evidence that he had ever bought any while living in Glasgow. The jury returned the unsatisfactory verdict, peculiar to Scots law, of: 'Not Proven'. Madeleine Smith was released and lived until 1926, when she died in New York at the age of 92.

..would certainly not have improved her complexion...

A POISONED AMBASSADOR

Another case of alleged arsenic poisoning was recalled by the death in 1987 of Mrs Clare Boothe Luce. She was US Ambassador to Italy from 1953 until her resignation through ill health in 1956. She complained first of tiredness, nervousness, nausea and loss of sensation in one foot. After treatment in the United States for anaemia she returned to duty, but became ill again. Her hair began to fall out, her nails became brittle and her mouth was inflamed. These symptoms suggested arsenic poisoning, which was said to have been confirmed by tests at the US Naval Hospital in Naples. Analysis of hair was obviously necessary; whether it was done is not known.

The secret service took over the investigation, but found no evidence of political poisoning. It was then suggested that flakes of green paint, containing lead arsenate, had been falling from the ceiling of the bedroom in the 17th century embassy which she used as a private office. Arsenic compounds were certainly used in green pigments in earlier times, though lead arsenate was known mainly as an insecticide and herbicide.

POISON GAS

The most poisonous arsenic compound is arsine (AsH_3), a gas with a garlic-like odour. Until the recent growth of the semiconductor industry, in which it has a number of uses, arsine was never made intentionally on a large scale but was produced only in chemical laboratories and during certain industrial processes. In 1908 Professor John Glaister of Glasgow University gave the first comprehensive review of arsine poisoning, describing 120 cases. Most of them resulted from the action of acids on metals, generating hydrogen which combined with arsenic present as an impurity. Sometimes arsine is released merely by the action of water on waste from metallurgical processes. Glaister's cases, many of them fatal, included 16 which occurred among military balloonists using hydrogen made by the action of acid on zinc, in which arsenic is a common impurity.

Glaister discussed the treatment and, more important, the prevention of arsine poisoning. More than half a century later, cases were still being brought to the attention of the second John Glaister (who succeeded his father in the chair of Forensic Medicine) and his colleague Hamilton Smith. The victims, some of whom died, were employed in cleaning tanks used in the manufacture of sulphuric acid by the lead chamber process (p 8) or in the recovery of non-ferrous metals from industrial waste.

An outbreak of arsine poisoning occurred as recently as 1974. The *Asiafreighter,* taking a mixed cargo from the United States to Europe, ran into heavy seas when nearing the British coast. The first mate and three of the crew went into a cargo hold to inspect the hull for damage. Shortly afterwards they became unwell. The next day they were all very ill and were airlifted to hospital, first in Truro and then in London. Investigation revealed many abnormalities including destruction of red blood cells (for which one patient required frequent transfusions), as well as kidney damage and skin eruptions. In a telephone consultation, Dr

Roy Goulding of the Poisons Unit at Guy's Hospital, made a confident diagnosis of arsine poisoning on the basis of the clinical findings. Only some days later was it found that the cargo included two cylinders of arsine, which were not mentioned in the ship's manifest; one cylinder had lost all of its contents through a valve damaged in the storm. The characteristic smell of arsine had not been masked by other pungent odours in the hold. Analysis made nine days after the accident showed an arsenic level of 236 PPM in the hair roots of the most seriously affected patient. Three months later, this patient's hair was showing 53 PPM of arsenic in a segment between three and four centimetres from the root.

Arsenic's bad reputation, both as a poison and, more recently, as a cause of cancer, is not altogether deserved. The biological effects of arsenic depend on the chemical form in which it is administered. The element itself is rarely prepared (because it has hardly any uses) but is not particularly toxic. The trioxide, (the substance colloquially known as arsenic) was favoured by the poisoners of earlier times mainly because it was easily and cheaply available and because it was difficult to detect in dead bodies. Organic compounds of arsenic are generally—with the notable exception of arsine much less toxic. In the long-running play *Arsenic and Old Lace,* the victims were disposed of by strychnine, which is certainly quicker and more reliable than arsenic as an instrument of homicide.

Arsenic was synonymous with poison in earlier times. In 1661 John Evelyn wrote, in his *Fumifugium,* a tract condemning air pollution:

> 'New Castle cole, as an expert Physician affirms, causeth consumptions, phthisicks, and the indisposition of the lungs, not only by the suffocating abundance of smoake, but also by its virulency: for all subterrany fuel hath a kind of virulent or arsenical vapour rising from it'.

Coal does contain a small amount of arsenic—about six PPM, a concentration too small for 17th century analysts to measure.

ARSENIC AND CANCER

Belief in arsenic as a cause of cancer was stimulated by John Paris, a physician who practised in England during the early years of the 19th century. His *Pharmacologia,* first published in 1812, became a popular textbook. Along with comments on the value of ale for dissolving kidney stones (not in the test tube, but in the patient) and the danger that turtle soup, if taken with tea, might turn to leather in the stomach, he described many cases of cancer among workers involved in the production of tin and copper and in farm animals grazing in the vicinity of the smelters. Paris attributed all of these diseases to poisoning by arsenic, which is certainly present in the ores of many metals. He dealt also with cancer of the scrotum, a disease which had been found by Percivall Pott, a London surgeon, to be particularly common among chimney sweeps. In his tract on the subject, published in 1775, Pott suggested that some constituent of the soot was responsible. Paris had no doubt that arsenic was to blame.

Paris's speculations on arsenical cancer in animals and human subjects living in the vicinity of smelters are still occasionally quoted with approval, but have not been confirmed. Many recent studies have given conflicting results. One typical investigation made in the United States reported, in 1975, elevated lung cancer rates in a number of counties where smelters were located. This finding was correct—but in several other counties containing smelters the lung cancer rate was below the national average. Occupational cancer of the scrotum, and of some other sites, is now known to be caused by exposure to organic compounds found in soot, tar, creosote and other materials derived from coal. There is no reliable evidence that cancer can be produced in animals by exposure to arsenic compounds, even in very large amounts. This situation is a reminder of the unreliability of animal studies as a guide to toxicity in human subjects.

It has been known for more than a century that consumption of arsenic (usually as the trioxide) over a long period of time occasionally leads to cancer of the skin. An interesting study bearing on this hazard was made in Glasgow in 1955. A man of 39 was referred to a dermatologist with a complaint of thickening of the skin of the palms and soles—a well known sign of chronic arsenical poisoning. It was found that the patient was under treatment for 'nervous debility' and had for 12 years been taking daily doses of a mixture containing arsenic trioxide, potassium bromide and phenobarbitone—a remarkable formulation designed to combine the qualities of a tonic and a sedative. He was admitted to hospital for further investigation. Neutron activation analysis performed by Professor Hamilton Smith showed arsenic levels as follows:

hair	65 PPM
skin	7
finger nail	11
beard hair	6
beard hair (8 days later)	2

Analysis of beard hair, taken with an electric razor, is a convenient way of indicating the daily removal of a toxic element.

The patient was advised to throw away his medicine and to abstain from arsenic for the rest of his days. He lived for several years, but eventually developed skin cancer which spread to the liver with fatal results.

Though the relationship of arsenic to skin cancer is well established, its ability to cause cancer of internal organs is uncertain. A publication from the Middlesex Hospital in London in 1963 reported six cases of lung cancer in patients who had been earlier treated with arsenic for several years. All showed skin changes characteristic of chronic arsenical poisoning. In five of the patients, these changes did not occur until many years after treatment had ceased. Three of the patients were moderate smokers, in whom the appearance of lung cancer may have been unrelated to arsenic, but the other three had never smoked. One of the non-smokers had been treated with arsenic up to the age of ten, but did not develop lung cancer (a rare disease in people who have never smoked) until 50 years later.

CHAPTER 7

The Chemist in the Witness Box

Trace element analysis has been used with varying degrees of success (and a few deplorable failures) in the investigation of crime. The scene of a crime always contains signs of the interaction between the criminal and the immediate environment. The criminal may leave something behind—for example, fingerprints, footprints, or hair, or may take something away—for example soil or bloodstains. The forensic scientist may be asked to measure and comment on the amount of a relevant substance found at the scene or on a suspect. More often the issue is identification rather than measurement: has a hair found at the scene come from the suspect?—has soil on the suspect's shoes come from the scene?—has a fragment of glass found on the suspect's jacket come from the broken window of a burgled building?—has a speck of paint found on the victim of a hit–and–run collision come from the suspect's car? To such questions the analyst cannot give a definite answer, but he can help the court by advising on the probability that two samples have a common source.

In the early days of neutron activation, it was hoped that trace element analysis of hair would be a valuable aid to justice, providing the chemical equivalent of a fingerprint. The analogy was not justified. Experience over nearly a century has established that a person's fingerprint is unique: the trace element composition of hair is not. It changes over a period of only a few days and, at any one time, is not the same for all of the hairs of a person's head. If the suspect is detained very soon after the crime, and if several hairs are taken from his head, something may be made of the trace element analysis; in other circumstances, apparent similarities in composition between hair found at the scene of the crime and hair taken from the suspect are of doubtful value. Evidence based on neutron activation analysis of hair has been presented in several trials in the United States. On some occasions the analytical work has been incompetently done by inexperienced enthusiasts, to the dismay of more conscientious exponents of the technique.

In many cases, the verdict does not seem to have been influenced by evidence from trace element analysis. But in other instances skilful advocates have, by attacking the competence of analysts, or the validity of novel techniques, succeeded in diverting attention from irrefutable evidence obtained by conventional enquiries. An example of this process was provided in the case of Marie Besnard.

THE BLACK WIDOW

The little town of Loudun in the west of France was the scene of an epidemic of demonic possession in the 17th century, described in Aldous Huxley's book *The Devils of Loudun*. Fifty years ago, one of the more colourful citizens was Marie Besnard, later known to her neighbours as 'The Black Widow'. In earlier times she would probably have been burnt as a witch; in 1952 she stood trial for the murder of a dozen people, including her father, her husband and relatives or friends whom she had tended during their last illnesses, and who had left her large sums of money. Remains of the deceased, analysed by traditional methods in the police laboratory in Marseilles, contained arsenic at several hundred times the normal level.

The prosecutor's evidence was formidable, and would probably have been decisive had not samples of exhumed hair been sent to Paris for examination by neutron activation analysis at the hands of Dr Henri Griffon, chief toxicologist in the police laboratory there. His findings were in line with those reported from Marseilles. The defence, unable to dispute the scientific evidence, admitted that the bodies of the twelve victims contained great quantities of arsenic—but attacked the accuracy of Griffon's analyses. They succeeded, with some difficulty, in identifying minor errors of technique and interpretation and in casting doubt on the validity of neutron activation analysis, which was then a very novel procedure.

The defence claimed also that arsenic found in exhumed bodies had come from the soil of the cemetery. This claim was not all plausible. Griffon analysed soil samples from the cemetery at Loudun. They contained arsenic (as all soil does) but most of it was insoluble and therefore not available for transfer to the internal organs of a buried corpse. In another experiment, locks of hair were buried for more than a year in the cemetery, at places where the concentration of arsenic in the soil was highest. The arsenic content of these samples was hardly changed. The defence did not dispute evidence that no arsenic was found in corpses unrelated to the Besnard case which had been buried in the same cemetery.

It is, however, often possible to find a scientist who supports a helpful hypothesis, however far-fetched it appears to the generality of qualified experts. So it was in the Besnard trial. Scientific witnesses suggested that arsenic could, through bacterial action, accumulate in a buried body to a level thousands of times higher than in the soil. This theoretical possibility was, by skilful pleading, made into a plausible defence and Marie Besnard was acquitted.

MURDER IN THE ARCTIC?

Hair analysis was used to investigate the alleged murder in 1871 of the American explorer Charles Francis Hall. The laboratory findings obtained nearly a century later pointed to arsenic poisoning, but would have been vigorously challenged, probably with success, had the case been tried in a court of law.

Hall was a celebrated polar explorer. He had led two expeditions in search of Sir John Franklin and his party, who set out in 1845 to find the Northwest passage and were never seen again (p 99). When the US Congress authorised an expedition to seek the North Pole, Hall was given command. The party set out in July, 1871, aboard the Polaris, a small steamship, and spent the winter of 1871–82 off the Greenland coast. Hall was not on good terms with the rest of the party. Emil Bessels, the expedition's naturalist and surgeon, treated him with contempt. Sidney Budington, the ship's captain, complained that Hall, who had rejected his advice to turn southward, was not competent to lead the expedition.

After a 14 day reconnaissance by sledge, Hall returned to the ship late in October, 1871. He drank a cup of coffee and immediately became severely ill. Two weeks later he died—according to Bessels of 'an apoplectical insult'. This diagnosis was supported by an official inquiry held in Washington in 1873, after the survivors of the expedition had been rescued.

Doubt remained, because Hall had often accused Bessels and others of trying to poison him. In August, 1968 Dr Franklin Paddock, a physician from Pittsfield, Massachusetts and Dr Chauncey Loomis, a professor of English at Dartmouth College in New Hampshire, flew to Polaris Promontory (where Hall had been buried) and performed an autopsy. Nothing remarkable was observed, but samples of soil, hair and finger nail were removed and sent to Toronto for examination by Dr Auskelis Perkons, a forensic scientist who is an expert in neutron activation analysis. The finger nail sample was 8 mm long, representing about 80 days' growth. Its arsenic content was 76.7 PPM in the first half millimetre from the root, decreasing fairly uniformly to 20.7 PPM at the other end. The hair, 2 cm long (corresponding to about 50 days growth), was also cut into sections for analysis. The arsenic concentration fell from 29.4 PPM in a 2.5 mm section nearest the root to 10.6 PPM in a 5 mm length at the tip.

These findings are consistent with a considerable intake of arsenic during a period of two or three months before Hall died. But this inference is undermined by some additional findings. Soil from the site contained 22 PPM of arsenic. This is about five times the normal level. Two samples of skull bone contained 10.3 and 11.7 PPM of arsenic; a normal level is about 0.05 PPM. These observations indicate heavy external contamination and (even though the hair and nail samples were washed before analysis) do not provide conclusive evidence for the cause of death. On the other hand the symptoms of Hall's illness, as reported by members of the expedition, were certainly consistent with arsenic poisoning. It is strange that this possibility was not regarded seriously by the official inquiry.

WHO FIRED THE GUN?

Poisoners are still at work—but, in many countries, the gun is now the commonest instrument of homicide. In the investigation of crimes involving firearms, it is often necessary to seek answers to two questions: (a) has the suspect fired a gun recently? and (b) have bullets found at the scene (or in the victim's body) been fired from a gun associated with the suspect?

When a criminal fires a gun, his purpose is to transfer a bullet to the victim. But tiny particles from the cartridge are inevitably transferred to his own person, from which they may be recovered and identified. The discharge of a bullet from a firearm is a two-stage process. When the trigger is pulled, the firing pin strikes the primer—a small quantity of a pressure-sensitive explosive, usually in a soft-metal capsule at the base of the cartridge. The primer explodes, sending a flame into the gunpowder which is the main propellant. The resulting explosion of the powder sends the bullet on its way. While this is happening, the barrel is filled with gases and particles. Most of this material escapes through the muzzle, but some is blown back onto the hand which has fired the weapon.

In earlier times, detection of traces of these materials was attempted by the paraffin test, in which molten paraffin wax was used to make a cast of the suspect's hand and then analysed for nitrates and nitrites, which are produced in the combustion of propellants. This test was notoriously unreliable and was eventually discredited because many courts would not accept it.

Since the mid-1960s a more acceptable technique has been developed by Professor Vincent Guinn and his colleagues, now at the University of California, Irvine. Almost all cartridge primers contain barium nitrate and antimony sulphide. Traces of barium and antimony are collected on a thin film of paraffin wax applied (in the molten state) to the back of the suspect's hand. The firing of one round from a 0.38 calibre revolver deposits about a microgram of barium and a few tenths of a microgram of antimony; these traces are readily measured by neutron activation analysis. The technique is sensitive enough to measure the very much smaller amounts of barium and antimony found on the hands of people who have not fired a gun.

Notwithstanding its extreme sensitivity, the analytical test for gunshot residue does not produce convincing evidence unless it is applied very soon after the weapon has been fired. Traces of antimony and barium disappear in about three hours of normal activity—and much more quickly if the hand is wiped or washed.

In this situation ballistic tests are sometimes useful. When a weapon is fired, random microscopic irregularities in the barrel produce a characteristic pattern of scratches on the bullet; no two barrels produce the same pattern. If a bullet is recovered in undistorted condition from the scene of a crime, and if the suspect weapon is also found, comparison of the scratch pattern can be made with a bullet fired from the weapon into a tank of water or a box of cotton wool. But if the suspect bullet is damaged, or if only fragments of it are found another method of identification must be used.

ANALYSIS OF BULLET LEAD

Neutron activation analysis has been developed, largely by Professor Guinn and his colleagues in California, into a powerful procedure for the investigation of crimes involving firearms. Bullets are made almost entirely of lead (an element for which activation analysis is not a particularly suitable technique) but other constituents can give a great deal of useful information. These minor elements are of two kinds. Some bullets are made from virtually pure lead. The refining process does not reach 100 percent purity, but leaves traces of copper, silver and arsenic. These elements can be estimated with good sensitivity by neutron activation analysis (p xi), without destroying or damaging the sample—an important consideration in criminal cases. Bullets are also made from lead which has been hardened by the inclusion of antimony at a concentration between 0.4 percent and 4 percent. Trace quantities of copper, silver and arsenic are, of course, found also in hardened lead. In Professor Guinn's laboratory the antimony, copper and silver contents of a fragment of no more than 10–20 mg of bullet lead can be measured in three minutes.

The lead supplier will usually prepare a substantial quantity—perhaps a ton or more—to the specification of the bullet manufacturer. A ton of lead can make (depending on the calibre) 50 000–100 000 bullets, closely similar in their contents of antimony (if specified) and of the three trace elements. The trace element content of the next batch will probably be different, since these elements are fortuitous impurities; their concentrations are not controlled by the lead supplier and are, indeed, of no interest to the bullet manufacturer since they are present in amounts too small to affect the properties of the ammunition.

To obtain useful information by neutron activation analysis, it is advantageous to have a bullet (or fragments of it) which may have been fired by the suspect, together with unused bullets from the suspect's weapon, or from some other source which can be associated with the suspect. When more than one weapon has been fired, analysis for trace elements may help to establish which gun was responsible for the death of a particular victim.

DEATH OF A PRESIDENT

John F Kennedy, 35th President of the United States, was assassinated in Dallas on 22 November, 1963. Soon afterwards. Lee Harvey Oswald was arrested a few miles away after shooting a police officer. He was suspected of the assassination, but never came to trial because he was himself killed, while in police custody, by a bystander.

The shots which killed the President and injured Governor John Connally, who was in the same car, appeared to come from the Texas School Book Depository. A search there discovered an Italian military rifle, three spent cartridge cases and, in the rifle, one unfired cartridge of the same make. The Warren Commission appointed to investigate the affair found that the rifle belonged to Oswald and that the three recovered cartridge cases had been fired from it. One bullet had missed its target completely and one had taken a complicated path, causing non-fatal wounds to the President and the Governor before coming to rest in the Governor's thigh and falling on to his stretcher. The third had killed the President, leaving fragments inside his skull and in the limousine.

Paraffin wax casts of Oswald's hand and right cheek, sent to Oak Ridge National Laboratory for neutron activation analysis to detect gunshot residues, were so carelessly taken and handled en route that the results of the tests were not useful. The bullet found on the Governor's stretcher was analysed by conventional techniques in the FBI laboratory, along with bullet fragments recovered from the limousine and the bodies of the victims. The results were inconclusive, but it was reported that the specimens were generally similar in chemical composition and could have been from the same brand of ammunition.

By then a rich crop of rumours had grown, suggesting an international conspiracy, involving two or more marksmen, firing from the same or different locations. For several years after 1964, eminent forensic scientists, as well as critics of the Warren Report, urged that the bullet lead specimens should be examined by a more sensitive technique, such as neutron activation analysis, to establish how many bullets had been fired, and whether they had all been fired by Oswald.

The Director of the FBI (J Edgar Hoover) resisted, with characteristic vigour, requests that bullets and fragments recovered after the assassination of the President should be subjected to neutron activation analysis. But in 1973 a letter dated 8 July 1964, from Hoover to the Warren Commission, was found in the National Archives. This letter disclosed—what Hoover and the Commission had concealed—that bullet lead specimens had indeed been subjected to activation analysis in May, 1964 by John F Gallagher, a forensic scientist on the staff of the FBI. The results were inconclusive. After a laborious legal struggle, Professor Guinn and Dr John Nichols, a forensic pathologist at the

Kansas Medical School, obtained in 1975 a copy of Gallagher's results—which, on first examination, did appear to be inconclusive.

But that was not the end of the story. Professor Guinn had by then analysed many samples of bullet lead from cartridges of the brand found in Oswald's rifle. He was surprised to find that they were, in trace element composition, different from almost every other specimen of bullet lead that he had studied. The concentrations of silver and particularly of antimony varied widely, not only from batch to batch, but even among the individual cartridges in a box of 20. It appeared that the bullets contained variable amounts of recycled lead. Guinn and Nichols persisted in urging that the bullet remnants recovered in Dallas should be re-examined by neutron activation analysis, using techniques which had advanced considerably since 1964.

Table 7.1 Silver and antimony (PPM ± standard deviation) in bullet lead recovered after the Kennedy assassination.

Sample number	Source	Silver	Antimony
Q1	Connally stretcher	8.8 ± 0.5	883 ± 9
Q2	fragment found in limousine	8.1 ± 0.6	602 ± 4
Q4, Q5	fragments from President's brain	7.9 ± 0.3	621 ± 4
Q9	fragment from Connally's wrist	9.8 ± 0.5	797 ± 7
Q14	fragments found in limousine	8.2 ± 0.4	642 ± 6

The analysis suggests clearly that samples Q1 and Q9 are from the same bullet and that Q2, Q4, Q5 and Q14 are from a different bullet.

In 1977, Professor Guinn was asked to make tests along these lines for the House of Representatives Select Committee on Assassinations. The tests were done in the nuclear reactor laboratory of the University of California, Irvine—under the constant surveillance of two armed Federal guards. The results are summarised in table 7.1. Coupled with other evidence they show that, to a high degree of probability:

(a) the recovered material came from only two bullets:
(b) one bullet took a complicated path, wounding the President and the

Governor in several places before coming to rest on the Governor's stretcher:

(c) the other bullet killed the President and broke into fragments inside his skull.

Further examination and rearrangement of Gallagher's results showed that they were consistent with these conclusions. Though uncertainty and controversy are rekindled from time to time, analytical chemistry has done much to refute claims that more than one gunman caused any of the injuries involved in the assassination.

WHO KILLED THE FIELD MARSHAL?

The Symbionese Liberation Army was a political organisation in California which probably never had more than a dozen members. Financial support for their aims was provided by robbing banks and stores—and by two million dollars extorted from the publisher William Randolph Hearst, after his daughter Patricia was kidnapped on 4 February 1974. Six members of the SLA, trapped in a house in Los Angeles on 17 May, 1974, died after a shoot-out with the police. During the action, and several thousand rounds of ammunition were fired and the house was destroyed by a fire which killed some of the occupants.

An autopsy showed that the leader of the SLA party, Donald DeFreeze (alias Field Marshal Cinque) had been killed by a single bullet, probably of 9 mm calibre. It was important to establish whether he had been shot by the police or had committed suicide. Fragments of metal were recovered from the wound track, but the bullet was not found. Professor Guinn was asked by the Coroner to analyse several specimens, including metal fragments from the wound, bullets of the three kinds found unused near DeFreeze's body and samples of 12 kinds of ammunition used by the police.

Analysis showed that the fragments found in DeFreeze's head contained only a few PPM of antimony, and had therefore come from a soft lead bullet. Since all of the police bullets, but only two of the three types of bullet used by DeFreeze, were made of lead hardened by antimony, it appeared that the Field Marshal had killed himself. This conclusion was undermined when a scientist in the Coroner's Office examined some of the fragments from the wound, using a scanning electron microscope. He found traces of lead (from the bullet) and brass (from the cartridge case), along with fragments of nickel-plated iron.

These must have come from the bullet jacket—but all of the bullets used by DeFreeze and by the police were in copper jackets. It was then discovered that one of the police officers in the siege had been using ammunition bought privately from a dealer in war surplus material. Thirteen types of 9 mm bullets which had been available in the United States after World War II were eventually found. One of them (made in Czechoslovakia in 1941) matched the trace element composition of the fragments recovered from DeFreeze's wound.

A CUNNING CRIMINAL

Neutron activation analysis of materials other than lead was used to good effect during a murder investigation in Syracuse, New York in 1970. The victim was a woman who had been strangled with a pair of tights. Clothing and bed linen were taken for forensic examination, but hair was not collected from the scene of the crime, because of a mistaken belief that it would be confused with dog hair which was present in large amounts; dog and human hair can readily be distinguished by microscopic examination.

Before long, the police identified a suspect, who was known to the victim and had a criminal record. He had learnt from experience and took care to leave no fingerprints, but was held on another charge, to allow time for further investigations—which were to give a striking demonstration of the value of trace element analysis. The suspect was employed in a factory dealing with tungsten carbide, sometimes compounded with graphite. The workers were inevitably contaminated by these materials. Samples of various tungsten carbide-graphite mixtures used as raw materials in the factory were found (table 7.2) to contain small amounts of cobalt and tantalum, along with larger amounts of tungsten. Other samples, taken at the scene of the crime, were analysed with the results shown in table 7.3. The minute amounts of titanium and antimony also found in samples of fabric from the victim's clothing were not significant, since these elements are commonly present in catalysts used in making synthetic fibres.

Tungsten, cobalt and tantalum are very unusual constituents of the normal domestic environment. The prosecutor suggested, on the basis of the evidence from neutron activation analysis, that these contaminants had been transferred from the clothing or body of a person who worked in an environment similar to that where the defendant was employed. This submission was accepted by the court and a conviction was obtained.

Table 7.2 Analysis of tungsten carbide-graphite samples.

	Tungsten (%)	Cobalt (%)	Tantalum (%)
Sample A	35	3.2	2.0
B	58	5.2	4.1
C	0.25	0.031	0.03

Table 7.3 Analysis of evidence material.

	Tungsten	Cobalt	Tantalum
victim's blouse,			
—stained area	315 PPM	9.26 PPM	9.89 PPM
—unstained area	31.7	ND	ND
victim's tights			
—top part of knot	61.4	2.73	1.14
—lower part of knot	28.7	1.18	0.37
suspect's hair	2244	39	15

ND—not detected

COUNTERFEIT SCOTCH

Trace element analysis, useful in the struggle against the moonshiners (p 99) achieved a notable success in exposing a sophisticated fraud involving liquor of very superior quality.

Johnnie Walker whisky, one of Scotland's most prestigious exports, was doing well in Brazil in 1971—indeed, all too well. The amount sold appeared to be more than was being imported. In Brazil, as in other countries, Scotch whisky commands a much higher price than the local product—so the loss of revenue was of great concern to the manufacturer's agent.

A search was made for evidence of smuggling or illicit distilling. Skilful police work uncovered a large quantity of whisky, claimed by the owner to be genuine Scotch. The boxes, bottles and labels all appeared authentic. The haul included also a stock of screwtops, cork rings and foil caps. The owner protested that this material had been legally acquired from a Brazilian manufacturer who supplied local distilleries. Samples of the items concerned were obtained from the local manufacturer and from the Scottish distillers. Further investigation was handed over to Dr Fausto Lima, head of the Radiochemistry Division of the Atomic Energy Institute in Sao Paulo.

Since moonshine whisky is always contaminated by toxic metals (p 99), samples of the genuine product and the suspected imitation were analysed. The results (table 7.4) were unexpected, since they showed no significant difference between the two samples. The seized whisky had obviously been made in a well-managed distillery and had, perhaps not surprisingly, brought no complaints from the customers who paid a high price for it. Activation analysis of cork rings showed no appreciable difference in trace element content. This finding was not significant, as it was afterwards found that almost all of the cork used in these items comes from the same source in Portugal.

Table 7.4 Brazilian whisky: trace element content, PPB.

	Imported	Apprehended
antimony	330	340
bromine	9	7
cadmium	60	40
chromium	16	27
cobalt	320	300
copper	200	130
gold	17	27
lead	1000	1000
manganese	9	12
scandium	8	5
zinc	17	22

The investigators then turned their attention to the foil caps used to seal (and decorate) the bottles. Foil used for this purpose is mostly lead, but contains small amounts of antimony and tin, as intentional constituents, along with minute amounts of several other elements present as unwanted impurities.

When foil caps from the various sources were analysed, the results (table 7.5) were decisive. It was apparent that the foil obtained from the seized bottles contained tin and antimony in about the same proportions as the locally-made foil and the unused foil seized along with the suspect bottles. The foil from bottles of genuine imported Johnnie Walker whisky had quite different proportions of tin and antimony. Analysis for other metals also showed a clear difference between the locally-made foil and the imported article. As the cases, bottles and labels were good enough to deceive the eye (and the whisky was good enough to deceive the Brazilian palate) it is doubtful whether the fraud would have been uncovered by any technique other than trace element analysis.

Table 7.5 Brazilian whisky: antimony and tin in bottle caps, PPM (mean of 10 samples).

	Source			
	A	B	C	D
antimony	2374	2381	2297	564
tin	4350	4186	3522	12694

A: caps from apprehended stock
B: caps from apprehended bottles
C: locally made caps
D: caps from imported bottles

FORGED PAINTINGS

The value of trace element analysis in the detection of non-violent crime has been demonstrated in many investigations related to art or archaeology.

The keeper of a museum or art gallery sometimes has to ask, of an object in his collection: where was it made? and: is it genuine or counterfeit? In the absence of an uninterrupted record of previous ownership, which can provide reassurance on both of these points, a subjective judgement may be based on visual inspection and professional experience. It is sometimes useful to augment this process by trace element analysis. The justification for this approach is as follows. The major constituents of materials such as paint, glass or coinage metal are normally controlled, explicitly or implicitly, so as to produce the properties desired for the uses to which they are put. Trace elements, present in such small amounts that they do not affect these properties, are not controlled. So the analysis of samples of paint, glass or coinage metal, made in different places or in different centuries, may show no difference in major constituents, but large differences in trace impurities. For this reason, the trace element content of a material can provide a useful indication, or confirmation, of its source.

Neutron activation analysis of a speck of paint from a picture will reveal the presence of many trace elements, present as unconsidered impurities in the white lead (lead hydroxycarbonate) which is commonly used, mixed with oil, to prime a canvas before painting. A study made in Holland in 1964 examined samples of white lead from Dutch and Flemish pictures painted since the 16th century. The trace element content of white lead (table 7.6) showed several changes during the past four centuries, reflecting developments in manufacturing practice. A similar study made in Germany at about the same time found differences in the concentrations of antimony, silver and manganese between Dutch and Venetian paintings of the 16th and 17th centuries. Investigations of this kind are obviously of potential value in the detection of counterfeit or false attribution of paintings—though such misdeeds are, for commercial reasons, seldom made public.

Table 7.6 Trace elements (PPM) in white lead used in Dutch paintings, 1500–1940.

	1500–1650	1650–1850	1850–1940	1940
silver	18–27	18–27	0–5	0–5
copper	150–220	150–220	0–60	0–60
mercury	3–7	3–7	0–1	0–1
manganese	70–110	70–110	0–12	0–12
chromium	225–500	0–35	0–35	0–35
zinc	0–6	0–60	0–60	600–6000
antimony	<0.01	<0.01	<0.01	30–110

(From Houtman J P W and Turkstra J 1964 *IAEA Symposium on Radiochemical Methods of Analysis* (Vienna: IAEA).)

An interesting technique, involving the activation of an entire painting, was developed by Dr Edward Sayre and his colleagues in New

York. The problem which inspired them was suspected forgery. Ralph Blakelock (1847–1919) was a self-taught American painter of modest fame, whose work became popular early in the present century and commanded exceptionally high prices. Many forgeries were made, in a labour-saving way. After Blakelock's productive period was ended by a mental breakdown in 1899, his daughter supported the family. She had been taught by her father and had absorbed much of his unusual style and techniques. Dealers found that much profit could be made by buying her paintings, removing her name, imitating her father's signature on the canvas and selling the forgeries at a high price.

Authentic Blakelock paintings and suspected forgeries were exposed to neutrons emerging from a nuclear reactor at Brookhaven National Laboratory. This reactor was designed for medical use and delivered a broad beam of neutrons. Under neutron bombardment, lead was not activated appreciably but measurable radioactivity was induced in many other materials. Two methods of assay were used. Firstly, an autoradiograph was made, in a darkroom, by placing a sheet of photographic film in close contact with the painting. This procedure was repeated at intervals during the first 24 hours and then after three days, a week and three weeks. Secondly, the radioactive emissions from the painting were examined at intervals.

For some time after the return of the painting from the reactor, the emitted radiation was dominated by gamma-rays of sodium–24 (half life 15 hours) and manganese–56 (half life 2.6 hours). One painting, believed to be a forgery, showed intense activity from aluminium–28, with half life 2.3 minutes. This activity was not detected in any of the genuine paintings. After a few days, the sodium activity had decayed considerably, leaving activity predominantly from arsenic and antimony; in later assays, mercury and cobalt were identified. Measurements made in these ways gave a mass of information which allowed a confident differentiation between authentic Blakelock paintings and forgeries. The autoradiograph gave clear evidence (which could not have been obtained in any other way) that the signature 'Marion Blakelock' had been partly removed and overpainted 'R A Blakelock'.

ANCIENT GLASS

The study of ancient glass (made between the second millenium BC and the 10th century AD) was until recently neglected by archaeologists, mainly because there appeared to be no identifiable physical or chemical properties by which the date and place of origin of a specimen could be estimated. Except for Islamic lead glass, probably made between the 8th and 10th centuries AD, all ancient glass is of the soda-lime-silica variety—commonly used today for windows and bottles—made by melting together sand, soda ash (impure sodium carbonate) and limestone. Small amounts of other materials were commonly added (for example, to remove coloration produced by traces of iron and other impurities) but their concentrations could not be accurately measured until the arrival of neutron activation analysis. An early application of this technique, using a nuclear reactor at the Brookhaven National Laboratory in the United States, examined several hundred specimens of ancient glass and identified five main groups with distinctive differences in chemical composition (table 7.7).

Table 7.7 Mean concentration of oxides in ancient glass.

	Oxide concentration (%)				
Type of glass	magnesium	potassium	manganese	antimony	lead
BC (second millenium)	3.6	1.13	0.032	0.058	0.0068
Antimony-rich	0.86	0.29	0.022	1.01	0.019
Roman	1.04	0.38	0.41	0.04	0.014
Early Islamic	4.9	1.41	0.47	0.021	0.0088
Islamic lead	0.33	0.026	0.022	0.081	36

(from Sayre E V and Smith R W 1962 *Science* **133** 1824–6)

ANALYSIS OF COINS

Ancient coins are of great interest to historians and archaeologists. Because of their intrinsic value, they have been well cared for by their original and subsequent owners. As the precious metals from which they are usually made are durable and resistant to corrosion, coins last longer than documents and longer than contemporary works of art in any other medium. They may carry the date of manufacture and can in any event often be dated by reference to the images and inscriptions that they bear. Coins record the growth and decay of dynasties and kingdoms and tell a great deal about religious cults, political movements, historical events and legends associated with their countries or cities of origin.

Although some 30 million ancient and medieval coins are in existence today, most of the information that they have provided has come from study of their external appearance. It is not surprising that only about 2000 such coins have been analysed by traditional chemical techniques, which are inevitably destructive.

The recent revolution in analytical chemistry has made it possible to discover, by non-destructive means, what a coin is made of. The relative proportions of various constituents can be found by streak analysis. In this technique, a small area of the edge of the coin is gently cleaned with emery paper, to remove any oxide layer, and then stroked with a piece of roughened quartz, to which a tiny streak, weighing perhaps 100 µg, is transferred. When subjected to neutron bombardment in a nuclear reactor, the quartz is not appreciably affected, but the streak yields radioactive isotopes of the metals composing it. Examination of the induced activity allows estimation of the relative proportions of the metals in the coin—without damaging it appreciably. Variations in gold or silver content can often be related to debasement of coinage for economic reasons—or enrichment in prosperous times.

Recent analyses have shown that many ancient silver coins contain a small amount (usually less than 1 percent) of gold as an impurity. The proportion of gold depended on the source of the silver. Streak analysis often allows the attribution of a coin, or series of coins, to a particular mint. Ancient coins offered for sale in the souks and bazaars of the Middle East are not all genuine. Modern fakes can readily be identified by streak analysis, because they are made from today's highly-purified silver which does not contain gold at a concentration characteristic of the genuine article.

SCIENTIFIC INSTRUMENTS

Antique scientific instruments have recently become interesting to collectors and, inevitably, to counterfeiters. In 1976, two Swiss chemists examined a portable sundial which, though bearing the signature 'Le Maire à Paris', used by an esteemed 18th century instrument maker, was thought to be of doubtful authenticity. Visual inspection showed no defects in design or craftsmanship. A small piece of brass from the object was subjected to neutron activation analysis, along with similar samples from 18th century instruments known to be genuine, and a piece of modern brass. The suspect article failed the test because—like the coins mentioned in the previous paragraph—it was too pure. The samples of 18th century brass contained indium, manganese and arsenic, which were not found in the modern brass or in the sample from the instrument under investigation.

DRAKE'S RELICS?

> 'In 38 deg. 30 min. we fell with a convenient and fit harborough, and June 17 came to anchor therein . . .'

Drake's Bay, some 30 miles north of San Francisco, is named after England's greatest navigator, who spent five weeks there in 1579, during his voyage round the world. The natives were most friendly and invested him with the symbols of kingship. Chaplain Fletcher, the chronicler of the voyage, recorded that

> 'before we went from thence, our Generall caused to be set vp a monument of our being there, as also of her maiesties and successors right and title to that kingdome; namely, a plate of brasse, fast nailed to a great and firme post; whereon is engrauen her graces name, and the day and yeare of our arriual there, and of the free giuing vp of the prouince and kingdome, both by the king and people, into her maiesties hands: together with her highnesse picture and armes, in a piece of sixpence currant English monie, shewing itself by a hole made of purpose through the plate; underneath was likewise engrauen the name of our Generall . . .'

In 1936 Mr Beryle Shinn, of Oakland, California, found a metal plate partly embedded in the soil near the head of San Quentin Bay, an arm of San Francisco Bay. He took the plate home, thinking that it might serve to repair a hole in his car. Some months later, when he started to use it for that purpose, he noticed that it bore some lettering. He scrubbed the plate but could not make much of the inscription. A friend advised him to take his find to Professor Herbert Bolton at the University of California, Berkeley. Bolton was a historian who had often told his students about Drake's Plate and suggested that they should look out for it when in the neighbourhood.

The plate found by Mr Shinn was crudely inscribed with the following text:

BE IT KNOWN VNTO ALL MEN BY THESE PRESENTS
IVNE 17 1579
BY THE GRACE OF GOD AND IN THE NAME OF HERR
MAIESTY QVEEN ELIZABETH OF ENGLAND AND HERR
SVCESSORS FOREVER I TAKE POSSESSION OF THIS
KINGDOME WHOSE KING AND PEOPLE FREELY RESIGNE
THEIR RIGHT AND TITLE IN THE WHOLE LAND VNTO HERR
MAIESTIES KEEPEING NOW NAMED BY MEE AND TO BEE
KNOWNE VNTO ALL MEN AS NOVA ALBION.
FRANCIS DRAKE

A roughly cut hole below the inscription presumably held the silver coin.

Bolton thought that the plate was genuine. It was bought for $3500 by a group of members of the California Historical Society and is now displayed in the Bancroft Library at Berkeley. The Society commissioned expert investigations, including trace element analysis, which suggested that the plate was several hundred years old. It had been shaped by hammering (rather than rolling) and was not homogeneous in chemical composition. It contained many trace impurities, including aluminium, antimony, calcium, manganese, silicon and tin, as well as the major constituents (copper and zinc) and small amounts of magnesium, iron and cadmium. The magnesium content of 102 PPM was much greater than in modern brass, where this element is not found at concentrations above 20 PPM. Other studies gave inconclusive or conflicting results; one suggested that the zinc content was too high for 16th century brass.

The style of lettering and the wording of the inscription have been criticised, mainly in unpublished papers, as not conforming to 16th century usage, though expert opinions are not unanimous on these points. Professor Samuel Eliot Morison, an eminent naval historian, declared that the plate was a hoax, comparable with the Piltdown Man. His opinion was somewhat undermined by the discovery that the illustration purporting to represent the plate in his book was actually a retouched photograph of a tinfoil replica made for the tourist trade by the McCoy Label Company of San Francisco.

DRAKE'S POTS?

The observation that potsherds (pieces of broken pottery) found in the vicinity bore some resemblance to 16th century Devonshire pottery led to speculation that they might have been left by Drake's party. Potters in the Devonshire towns of Bideford and Barnstable (which are not far from Drake's home port of Plymouth) were well known for their earthenware during the 16th century. Samples of their work were collected by Dr Jacqueline Olin of the Smithsonian Institution and Dr Edward Sayre of Brookhaven National Laboratory, along with sherds excavated from various American sites and believed to have originated in north Devon.

Small samples, about a square centimetre in area and a millimetre in thickness, were cut from each sherd. The concentrations of iron (about 4 percent) and of seven trace elements were estimated by activation analysis, using one of the reactors at Brookhaven. The results, published in 1971, suggested that the fragments found at Drake's Bay had not originated in North Devon.

Postscript: Sweeping up the Crumbs

Our journey has taken us into many byways—and a few highways—of industry, literature, medicine, art, history, crime: into the microscopic environment of cells and enzymes, into the vast emptiness of the cosmos. The study of trace elements reminds us of the essential coherence of these diverse realms, linked as they are by the economy of design and purpose so conspicuous in the Creator's strategy.

Because the trace elements constitute a scarce resource, they have to fill many roles—toxic, beneficial, essential, fortuitous—which sometimes appear to conflict or to overlap. Mechanisms for their deployment and conservation have many interesting features. Some essential trace elements, for example, are washed from the land into rivers and eventually into the sea. They might be expected to remain for a long time in solution or in the ocean sediments, to the detriment of land plants and animals which need them—especially since the losses are sometimes inexplicably larger than expected. It was shown during the 1960s that the amount of sulphur washed into the sea was much greater than could be provided by the known sources, such as weathering of rocks, burning of fossil fuels and extraction from the soil by plants. Since life continues on the Earth, these losses are made good—but how?

The answer appears to lie in seaweed and other algae. Some of these organisms have a remarkable capacity for extracting sulphur from sea water and using it to produce dimethyl sulphide, a volatile substance which is then carried back to the land by the wind. Another variety of seaweed concentrates iodine to make methyl iodide and in this way restores to the land iodine which would otherwise be lost.

Methylation is a process widely found in the living world. Bacteria in the sea bed sediments convert unwanted mercury, lead, arsenic and possibly selenium into volatile methyl compounds which ascend to the surface. They are then decomposed in the atmosphere and returned to the land. Methyl mercury is potentially toxic (p 84) but is sufficiently diluted in the sea to be harmless to fish—and, in general, to people who eat fish which has not been exposed to additional man-made sources. In these ways potentially toxic trace elements are kept on the move, preventing an unwelcome build-up in sensitive places, and the limited supply of some essential elements is continually recycled.

The remarkable influence of minute amounts of trace elements in living systems reminds us of a strategic principle which contributes greatly to the efficiency and economy of the living world. This principle is demonstrated by the uneven distribution of qualities. In human societies there are a few people with exceptional mental or physical endowments, which allow them to take on specialised roles in the community, a large number clustered around the norm and a minority with conspicuous deficiencies. A community in which everyone was equally intelligent, equally brave and equally aggressive would not be successful. Uniformity breeds mediocrity.

Among the chemical elements some are abundant and versatile, appearing in a wide range of compounds and some, as we have seen in the preceding pages, are scarce and highly specialised. The contrast between abundance and scarcity is evident also in a wider context. The universe consists mainly of hydrogen and helium, with not much of anything else. Elements heavier than helium are the products of the primeval thermonuclear reactor or of subsequent supernova explosions. Stars are accidental condensations of space dust. Solar systems are, on the cosmic scale, very rare. Abodes of life—particularly of intelligent introspective life—are curious anomalies. We ourselves are no more than traces—the mere crumbs of creation.

Further reading

CHAPTER 1

Forshufvud S 1972 *Who Killed Napoleon?* (London: Hutchinson)
Richardson F M 1974 *Napoleon's Death—An Inquest* (London: Kimber)
Valkovic V 1977 *Trace Elements in Human Hair* (New York: Garland STPM Press)
Weider B and Hapgood D 1983 *The Murder of Napoleon* (London: Transworld)

CHAPTER 2

Davies B E (ed) 1980 *Applied Soil Trace Elements* (New York: Wiley)
Desmond A J 1975 *The Hot-blooded Dinosaurs* (London: Blond and Briggs)
Hopps H C and Cannon H L (eds) 1972 Geochemical Environment in Relation to Health and Disease *Ann. NY Acad. Sci.* **199**
Lacey R F 1981 *Technical Report TR 171—Changes in Water Hardness and Cardiovascular Death-rates* (Medmenham: Water Research Centre)
Powell P *et al* 1982 *Technical Report TR 178—Water Quality and Cardiovascular Disease in British Towns* (Medmenham: Water Research Centre)
Thornton I (ed) 1983 *Applied Environmental Geochemistry* (New York: Academic Press)

CHAPTER 3

Bryce-Smith D and Hodgkinson L 1986 *The Zinc Solution* (London: Arrow Books)
Davies I J T 1972 *The Clinical Significance of the Essential Biological Metals* (London: Heinemann)
Levander O A and Cheng L (eds) 1980 Micronutrient Interactions: Vitamins,

Minerals and Hazardous Elements *Ann. NY Acad. Sci.* **355**
Recommended Dietary Allowances 1980 (Washington Academy of Sciences)
Paul A A and Southgate D A T 1978 *McCance and Widdowson's—The Composition of Foods* (London: HMSO)
Passwater R A 1980 *Selenium as Food and Medicine* (New Canaan: Keats)
Schroeder H A 1976 *Trace Elements and Nutrition* (London: Faber)
Sunderman F W (ed) 1984 *Nickel in the Human Environment* (Lyon: IARC)
Tucker A 1972 *The Toxic Metals* (London: Pan/Ballantine)
Underwood E J 1977 *Trace Elements in Human and Animal Nutrition* fourth edition (New York: Academic Press) (The fifth edition of this book is a multi-author work, edited by W Mertz. Volume 2 was published in 1986; volume 1 is to appear later)

CHAPTER 4

Dewhurst K 1957 *The Quicksilver Doctor: the Life and Times of Thomas Dover, Physician and Adventurer* (Bristol: Wright)
Goldwater L 1972 *Mercury: A History of Quicksilver* (Baltimore: York Press)

CHAPTER 5

Beattie O and Geiger J 1987 *Frozen in Time: the Fate of the Franklin Expedition* (London: Bloomsbury)
Dana S L 1848 *Lead Diseases* (London: Daniel Bixby)
Nriagu J O 1983 *Lead and Lead Poisoning in Antiquity* (Chichester: Wiley)
Royal Commission on Environmental Pollution 1986 *Ninth Report—Lead in the Environment* (London: HMSO)
Rutter M and Jones R J 1983 *Lead versus Health* (Chichester: Wiley)

CHAPTER 6

Buchanan G S 1901 *Report to the Local Government Board on Recent Epidemic Arsenical Poisoning Attributed to Beer* (London: HMSO)
Buchanan W D 1962 *Toxicity of Arsenic Compounds* (London: Elsevier)
Hunt P de V 1965 *The Madeleine Smith Affair* (London: Collier-Macmillan)
Jesse F T (ed) 1927 *The Trial of Madeleine Smith* (London: Hodge)
Royal Commission on Arsenical Poisoning First report, 1901; Final Report, 1903 (London: HMSO)

CHAPTER 7

Besnard M 1963 *The Trial of Marie Besnard* (London: Heinemann)

Brill R H (ed) 1971 *Science and Archaeology* (Cambridge: MIT Press)

Guinn V P 1979 JFK Assassination: Bullet Analysis *Analytical Chemistry* **51 (4)** 484A–493A

Karjala D S 1971 The Evidentiary Uses of Neutron Activation Analysis *California Law Review* **59** 997–1080

Morison S E 1974 *The European Discovery of America: The Southern Voyages 1497–1616* (Oxford: Oxford University Press)

Thorwald J 1966 *Proof of Poison* (London: Thames and Hudson)

——1967 *Crime and Science* (New York: Harcourt Brace)

APPENDIX

Elements of natural occurrence

Name	Symbol	Atomic number	Name	Symbol	Atomic number
actinium	Ac	89	erbium	Er	68
aluminium	Al	13	europium	Eu	63
antimony	Sb	51	fluorine	F	9
argon	Ar	18	francium	Fr	87
arsenic	As	33			
barium	Ba	56	gadolinium	Gd	64
beryllium	Be	4	gallium	Ga	31
bismuth	Bi	83	germanium	Ge	32
boron	B	5	gold	Au	79
bromine	Br	35	hafnium	Hf	72
cadmium	Cd	48	helium	He	2
caesium	Cs	55	holmium	Ho	67
calcium	Ca	20	hydrogen	H	1
carbon	C	6	indium	In	49
cerium	Ce	58	iodine	I	53
chlorine	Cl	17	iridium	Ir	77
chromium	Cr	24	iron	Fe	26
cobalt	Co	27	krypton	Kr	36
copper	Cu	29	lanthanum	La	57
dysprosium	Dy	66	lead	Pb	82

Name	Symbol	Atomic number	Name	Symbol	Atomic number
lithium	Li	3	ruthenium	Ru	44
lutetium	Lu	71	samarium	Sm	62
magnesium	Mg	12	scandium	Sc	21
manganese	Mn	25	selenium	Se	34
mercury	Hg	80	silicon	Si	14
molybdenum	Mo	42	silver	Ag	47
neodymium	Nd	60	sodium	Na	11
neon	Ne	10	strontium	Sr	38
nickel	Ni	28	sulphur	S	16
niobium	Nb	41	tantalum	Ta	73
nitrogen	N	7	tellurium	Te	52
osmium	Os	76	terbium	Tb	65
oxygen	O	8	thallium	Tl	81
			thorium	Th	90
palladium	Pd	46	thulium	Tm	69
phosphorus	P	15	tin	Sn	50
platinum	Pt	78	titanium	Ti	22
polonium	Po	84	tungsten	W	74
potassium	K	19	uranium	U	92
praseodymium	Pr	59	vanadium	V	23
promethium	Pm	61	xenon	Xe	54
protoactinium	Pa	91			
radium	Ra	88	ytterbium	Yb	70
radon	Rn	86	yttrium	Y	39
rhenium	Re	75	zinc	Zn	30
rhodium	Rh	45	zirconium	Zr	40
rubidium	Rb	37			

Index